Qiguo Zhao
Jikun Huang

Agricultural Science & Technology in China: A Roadmap to 2050

Qiguo Zhao
Jikun Huang

Agricultural Science & Technology in China: A Roadmap to 2050

With 13 figures

Editors

Qiguo Zhao

Institute of Soil Science, CAS
210008, Nanjing, China
E-mail: qgzhao@issas.ac.cn

Jikun Huang

Center for Chinese Agricultural Policy
Institute of Geographic Sciences and Natural Resources Research, CAS
100101, Beijing, China
E-mail: jkhuang.ccap@igsnrr.ac.cn

ISBN 978-7-03-029978-9
Science Press Beijing

ISBN 978-3-642-19127-5 e-ISBN 978-3-642-19128-2
Springer Heidelberg Dordrecht London New York

Cover design: Frido Steinen-Broo, EStudio Calamar, Spain

Printed on acid-free paper

Springer is part of Springer Science+Business Media (www.springer.com)

Members of the Editorial Committee and the Editorial Office

Research Group on Agriculture of the Chinese Academy of Sciences

Adviser: Zhensheng Li

Head: Qiguo Zhao

Deputy Head: Jikun Huang

Members:

Xiangdong Deng	Institute of Genetics and Developmental Biology, CAS
Ziyuan Duan	Bureau of Life Sciences and Biotechnology, CAS
Jianfang Gui	Institute of Hydrobiology, CAS
Jikun Huang	Center for Chinese Agricultural Policy, Institute of Geographic Sciences and Natural Resources Research, CAS
Gongshe Liu	Institute of Botany, CAS
Leqing Qu	Institute of Botany, CAS
Qianhua Shen	Institute of Genetics and Developmental Biology, CAS
Bo Sun	Institute of Soil Science, CAS
Hong Tan	Chengdu Institute of Biology, CAS
Daowen Wang	Institute of Genetics and Developmental Biology, CAS
Jianhai Xiang	Institute of Oceanology, CAS
Zhigang Xu	Center for Chinese Agricultural Policy, Institute of Geographic Sciences and Natural Resources Research, CAS
Jianxia Yuan	National Science Library, CAS
Aimin Zhang	Institute of Genetics and Developmental Biology, CAS
Jiabao Zhang	Institute of Soil Science, CAS
Linxiu Zhang	Center for Chinese Agricultural Policy, Institute of Geographic Sciences and Natural Resources Research, CAS
Chunjiang Zhao	National Engineering Research Center for Information Technology in Agriculture
Qiguo Zhao	Institute of Soil Science, CAS

Foreword to the Roadmaps 2050*

China's modernization is viewed as a transformative revolution in the human history of modernization. As such, the Chinese Academy of Sciences (CAS) decided to give higher priority to the research on the science and technology (S&T) roadmap for priority areas in China's modernization process. What is the purpose? And why is it? Is it a must? I think those are substantial and significant questions to start things forward.

Significance of the Research on China's S&T Roadmap to 2050

We are aware that the National Mid- and Long-term S&T Plan to 2020 has already been formed after two years' hard work by a panel of over 2000 experts and scholars brought together from all over China, chaired by Premier Wen Jiabao. This clearly shows that China has already had its S&T blueprint to 2020. Then, why did CAS conduct this research on China's S&T roadmap to 2050?

In the summer of 2007 when CAS was working out its future strategic priorities for S&T development, it realized that some issues, such as energy, must be addressed with a long-term view. As a matter of fact, some strategic researches have been conducted, over the last 15 years, on energy, but mainly on how to best use of coal, how to best exploit both domestic and international oil and gas resources, and how to develop nuclear energy in a discreet way. Renewable energy was, of course, included but only as a supplementary energy. It was not yet thought as a supporting leg for future energy development. However, greenhouse gas emissions are becoming a major world concern over

* It is adapted from a speech by President Yongxiang Lu at the First High-level Workshop on China's S&T Roadmap for Priority Areas to 2050, organized by the Chinese Academy of Sciences, in October, 2007.

the years, and how to address the global climate change has been on the agenda. In fact, what is really behind is the concern for energy structure, which makes us realize that fossil energy must be used cleanly and efficiently in order to reduce its impact on the environment. However, fossil energy is, pessimistically speaking, expected to be used up within about 100 years, or optimistically speaking, within about 200 years. Oil and gas resources may be among the first to be exhausted, and then coal resources follow. When this happens, human beings will have to refer to renewable energy as its major energy, while nuclear energy as a supplementary one. Under this situation, governments of the world are taking preparatory efforts in this regard, with Europe taking the lead and the USA shifting to take a more positive attitude, as evidenced in that: while fossil energy has been taken the best use of, renewable energy has been greatly developed, and the R&D of advanced nuclear energy has been reinforced with the objective of being eventually transformed into renewable energy. The process may last 50 to 100 years or so. Hence, many S&T problems may come around. In the field of basic research, for example, research will be conducted by physicists, chemists and biologists on the new generation of photovoltaic cell, dye-sensitized solar cells (DSC), high-efficient photochemical catalysis and storage, and efficient photosynthetic species, or high-efficient photosynthetic species produced by gene engineering which are free from land and water demands compared with food and oil crops, and can be grown on hillside, saline lands and semi-arid places, producing the energy that fits humanity. In the meantime, although the existing energy system is comparatively stable, future energy structure is likely to change into an unstable system. Presumably, dispersive energy system as well as higher-efficient direct current transmission and storage technology will be developed, so will be the safe and reliable control of network, and the capture, storage, transfer and use of CO_2, all of which involve S&T problems in almost all scientific disciplines. Therefore, it is natural that energy problems may bring out both basic and applied research, and may eventually lead to comprehensive structural changes. And this may last for 50 to 100 years or so. Taking the nuclear energy as an example, it usually takes about 20 years or more from its initial plan to key technology breakthroughs, so does the subsequent massive application and commercialization. If we lose the opportunity to make foresighted arrangements, we will be lagging far behind in the future. France has already worked out the roadmap to 2040 and 2050 respectively for the development of the 3rd and 4th generation of nuclear fission reactors, while China has not yet taken any serious actions. Under this circumstance, it is now time for CAS to take the issue seriously, for the sake of national interests, and to start conducting a foresighted research in this regard.

This strategic research covers over some dozens of areas with a long-term view. Taking agriculture as an example, our concern used to be limited only to the increased production of high-quality food grains and agricultural by-products. However, in the future, the main concern will definitely be given to the water-saving and ecological agriculture. As China is vast in territory, diversified technologies in this regard are the appropriate solutions. Animal husbandry has been used by developed countries, such as Japan and Denmark, to make bioreactor and pesticide as well. Plants have been used by Japan to make bioreactors which are safer and cost-effective than that made from animals. Potato, strawberry, tomato and the like have been bred in germ-free greenhouses, and value-added products have been made through gene transplantation technology. Agriculture in China must not only address the food demands from its one billions-plus population, but also take into consideration of the value-added agriculture by-products and the high-tech development of agriculture as well. Agriculture in the future is expected to bring out some energies and fuels needed by both industry and man's livelihood as well. Some developed countries have taken an earlier start to conduct foresighted research in this regard, while we have not yet taken sufficient consideration.

Population is another problem. It will be most likely that China's population will not drop to about 1 billion until the end of this century, given that the past mistakes of China's population policy be rectified. But the subsequent problem of ageing could only be sorted out until the next century. The current population and health policies face many challenges, such as, how to ensure that the 1.3 to 1.5 billion people enjoy fair and basic public healthcare; the necessity to develop advanced and public healthcare and treatment technologies; and the change of research priority to chronic diseases from infectious diseases, as developed countries have already started research in this regard under the increasing social and environmental change. There are many such research problems yet to be sorted out by starting from the basic research, and subsequent policies within the next 50 years are in need to be worked out.

Space and oceans provide humanity with important resources for future development. In terms of space research, the well-known Manned Spacecraft Program and China's Lunar Exploration Program will last for 20 or 25 years. But what will be the whole plan for China's space technology? What is the objective? Will it just follow the suit of developed countries? It is worth doing serious study in this regard. The present spacecraft is mainly sent into space with chemical fuel propellant rocket. Will this traditional propellant still be used in future deep space exploration? Or other new technologies such as electrical propellant, nuclear energy propellant, and solar sail technologies be developed? We haven't yet done any strategic research over these issues, not even worked out any plans. The ocean is abundant in mineral resources, oil and gas, natural gas hydrate, biological resources, energy and photo-free biological evolution, which may arise our scientific interests. At present, many countries have worked out new strategic marine plans. Russia, Canada, the USA, Sweden and Norway have centered their contention upon the North Pole, an area of strategic significance. For this, however, we have only limited plans.

The national and public security develops with time, and covers both conventional and non-conventional security. Conventional security threats only refer to foreign invasion and warfare, while, the present security threat may come out from any of the natural, man-made, external, interior, ecological, environmental, and the emerging networking (including both real and virtual) factors. The conflicts out of these must be analyzed from the perspective of human civilization, and be sorted out in a scientific manner. Efforts must be made to root out the cause of the threats, while human life must be treasured at any time.

In general, it is necessary to conduct this strategic research in view of the future development of China and mankind as well. The past 250 years' industrialization has resulted in the modernization and better-off life of less than 1 billion people, predominantly in Europe, North America, Japan and Singapore. The next 50 years' modernization drive will definitely lead to a better-off life for 2–3 billion people, including over 1 billion Chinese, doubling or tripling the economic increase over that of the past 250 years, which will, on the one hand, bring vigor and vitality to the world, and, on the other hand, inevitably challenge the limited resources and eco-environment on the earth. New development mode must be shaped so that everyone on the earth will be able to enjoy fairly the achievements of modern civilization. Achieving this requires us, in the process of China's modernization, to have a foresighted overview on the future development of world science and human civilization, and on how science and technology could serve the modernization drive. S&T roadmap for priority areas to 2050 must be worked out, and solutions to core science problems and key technology problems must be straightened out, which will eventually provide consultations for the nation's S&T decision-making.

Possibility of Working out China's S&T Roadmap to 2050

Some people held the view that science is hard to be predicted as it happens unexpectedly and mainly comes out of scientists' innovative thinking, while, technology might be predicted but at the maximum of 15 years. In my view, however, S&T foresight in some areas seems feasible. For instance, with the exhaustion of fossil energy, some smart people may think of transforming solar energy into energy-intensive biomass through improved high-efficient solar thin-film materials and devices, or even developing new substitute. As is driven by huge demands, many investments will go to this emerging area. It is, therefore, able to predict that, in the next 50 years, some breakthroughs will undoubtedly be made in the areas of renewable energy and nuclear energy as well. In terms of solar energy, for example, the improvement of photoelectric conversion efficiency and photothermal conversion efficiency will be the focus. Of course, the concrete technological solutions may be varied, for example, by changing the morphology of the surface of solar cells and through the reflection, the entire spectrum can be absorbed more efficiently; by developing multi-layer functional thin-films for transmission and absorption; or by introducing of nanotechnology and quantum control technology, etc. Quantum control research used to limit mainly to the solution to information functional materials. This is surely too narrow. In the future, this research is expected to be extended to the energy issue or energy-based basic research in cutting-edge areas.

In terms of computing science, we must be confident to forecast its future development instead of simply following suit as we used to. This is a possibility rather than wild fancies. Information scientists, physicists and biologists could be engaged in the forward-looking research. In 2007, the Nobel Physics Prize was awarded to the discovery of colossal magneto-resistance, which was, however, made some 20 years ago. Today, this technology has already been applied to hard disk store. Our conclusion made, at this stage, is that: it is possible to make long-term and unconventional S&T predictions, and so is it to work out China's S&T roadmap in view of long-term strategies, for example, by 2020 as the first step, by 2030 or 2035 as the second step, and by 2050 as the maximum.

This possibility may also apply to other areas of research. The point is to emancipate the mind and respect objective laws rather than indulging in wild fancies. We attribute our success today to the guidelines of emancipating the mind and seeking the truth from the facts set by the Third Plenary Session of the 11th Central Committee of the Communist Party of China in 1979. We must break the conventional barriers and find a way of development fitting into China's reality. The history of science tells us that discoveries and breakthroughs could only be made when you open up your mind, break the conventional barriers, and make foresighted plans. Top-down guidance on research with increased financial support and involvement of a wider range of talented scientists is not in conflict with demand-driven research and free discovery of science as well.

Necessity of CAS Research on China's S&T Roadmap to 2050

Why does CAS launch this research? As is known, CAS is the nation's highest academic institution in natural sciences. It targets at making basic, forward-looking and strategic research and playing a leading role in China's science. As such, how can it achieve this if without a foresighted view on science and technology? From the perspective of CAS, it is obligatory to think, with a global view, about what to do after the 3rd Phase of the Knowledge Innovation Program (KIP). Shall we follow the way as it used to? Or shall we, with a view of national interests, present our in-depth insights into different research disciplines, and make efforts to reform the organizational structure and system, so that the innovation capability of CAS and the nation's science and technology mission will be raised to a new height? Clearly, the latter is more positive. World science and technology develops at a lightening speed. As global economy grows, we are aware that we will be lagging far behind if without making progress, and will lose the opportunity if without making foresighted plans. S&T innovation requires us to make joint efforts, break the conventional barriers and emancipate the mind. This is also what we need for further development.

The roadmap must be targeted at the national level so that the strategic research reports will form an important part of the national long-term program. CAS may not be able to fulfill all the objectives in the reports. However, it can select what is able to do and make foresighted plans, which will eventually help shape the post-2010 research priorities of CAS and the guidelines for its future reform.

Once the long-term roadmap and its objectives are identified, system mechanism, human resources, funding and allocation should be ensured for full implementation. We will make further studies to figure out: What will happen to world innovation system within the next 30 to 50 years? Will universities, research institutions and enterprises still be included in the system? Will research institutes become grid structure? When the cutting-edge research combines basic science and high-tech and the transformative research integrates the cutting-edge research with industrialization, will that be the research trend in some disciplines? What will be the changes for personnel structure, motivation mechanism and upgrading mechanism within the innovation system? Will there be any changes for the input and structure of innovation resources? If we could have a clear mind of all the questions, make foresighted plans and then dare to try out in relevant CAS institutes, we will be able to pave a way for a more competitive and smooth development.

Social changes are without limit, so are the development of science and technology, and innovation system and management as well. CAS must keep moving ahead to make foresighted plans not only for science and technology, but also for its organizational structure, human resources, management modes, and resource structures. By doing so, CAS will keep standing at the forefront of science and playing a leading role in the national innovation system, and even, frankly speaking, taking the lead in some research disciplines in the world. This is, in fact, our purpose of conducting the strategic research on China's S&T roadmap.

Prof. Dr.-Ing. Yongxiang Lu

President of the Chinese Academy of Sciences

Preface to the Roadmaps 2050

CAS is the nation's think tank for science. Its major responsibility is to provide S&T consultations for the nation's decision-makings and to take the lead in the nation's S&T development.

In July, 2007, President Yongxiang Lu made the following remarks: "In order to carry out the Scientific Outlook of Development through innovation, further strategic research should be done to lay out a S&T roadmap for the next 20–30 years and key S&T innovation disciplines. And relevant workshops should be organized with the participation of scientists both within CAS and outside to further discuss the research priorities and objectives. We should no longer confine ourselves to the free discovery of science, the quantity and quality of scientific papers, nor should we satisfy ourselves simply with the Principal Investigators system of research. Research should be conducted to address the needs of both the nation and society, in particular, the continued growth of economy and national competitiveness, the development of social harmony, and the sustainability between man and nature. "

According to the Executive Management Committee of CAS in July, 2007, CAS strategic research on S&T roadmap for future development should be conducted to orchestrate the needs of both the nation and society, and target at the three objectives: the growth of economy and national competitiveness, the development of social harmony, and the sustainability between man and nature.

In August, 2007, President Yongxiang Lu further put it: "Strategic research requires a forward-looking view over the world, China, and science & technology in 2050. Firstly, in terms of the world in 2050, we should be able to study the perspectives of economy, society, national security, eco-environment, and science & technology, specifically in such scientific disciplines as energy, resources, population, health, information, security, eco-environment, space and oceans. And we should be aware of where the opportunities and challenges lie. Secondly, in terms of China's economy and society in 2050, we should take into consideration of factors like: objectives, methods, and scientific supports needed for economic structure, social development, energy structure, population and health, eco-environment, national security and innovation capability. Thirdly, in terms of the guidance of Scientific Outlook of Development on science and technology, it emphasizes the people's interests and development, science and technology, science and economy, science and society, science and eco-environment, science and culture, innovation and collaborative development. Fourthly, in terms of the supporting role of research in scientific development, this includes how to optimize the economic structure and boost economy, agricultural development, energy structure, resource conservation, recycling economy, knowledge-based society, harmonious coexistence between man and nature, balance of regional development, social harmony, national security, and international cooperation. Based on these, the role of CAS will be further identified."

Subsequently, CAS launched its strategic research on the roadmap for priority areas to 2050, which comes into eighteen categories including: energy, water resources, mineral resources, marine resources, oil and gas, population and health, agriculture, eco-environment, biomass resources, regional development, space, information, advanced manufacturing, advanced materials, nano-science, big science facilities, cross-disciplinary and frontier research, and national and public security. Over 300 CAS experts in science, technology, management and documentation & information, including about 60 CAS members, from over 80 CAS institutes joined this research.

Over one year's hard work, substantial progress has been made in each research group of the scientific disciplines. The strategic demands on priority areas in China's modernization drive to 2050 have been strengthened out; some core science problems and key technology problems been set forth; a relevant S&T roadmap been worked out based on China's reality; and eventually the strategic reports on China's S&T roadmap for eighteen priority areas to 2050 been formed. Under the circumstance, both the Editorial Committee and Writing Group, chaired by President Yongxiang Lu, have finalized the general report. The research reports are to be published in the form of CAS strategic research serial reports, entitled *Science and Technology Roadmap to China 2050: Strategic Reports of the Chinese Academy of Sciences.*

The unique feature of this strategic research is its use of S&T roadmap approach. S&T roadmap differs from the commonly used planning and technology foresight in that it includes science and technology needed for the future, the roadmap to reach the objectives, description of environmental changes, research needs, technology trends, and innovation and technology development. Scientific planning in the form of roadmap will have a clearer scientific objective, form closer links with the market, projects selected be more interactive and systematic, the solutions to the objective be defined, and the plan be more feasible. In addition, by drawing from both the foreign experience on roadmap research and domestic experience on strategic planning, we have formed our own ways of making S&T roadmap in priority areas as follows:

(1) Establishment of organization mechanism for strategic research on S&T roadmap for priority areas

The Editorial Committee is set up with the head of President Yongxiang Lu and the involvement of Chunli Bai, Erwei Shi, Xin Fang, Zhigang Li, Xiaoye Cao and Jiaofeng Pan. And the Writing Group was organized to take responsibility of the research and writing of the general report. CAS Bureau of Planning and Strategy, as the executive unit, coordinates the research, selects the scholars, identifies concrete steps and task requirements, sets forth research approaches, and organizes workshops and independent peer reviews of the research, in order to ensure the smooth progress of the strategic research on the S&T roadmap for priority areas.

(2) Setting up principles for the S&T roadmap for priority areas

The framework of roadmap research should be targeted at the national level, and divided into three steps as immediate-term (by 2020), mid-term (by 2030) and long-term (by 2050). It should cover the description of job requirements, objectives, specific tasks, research approaches, and highlight core science problems and key technology problems, which must be, in general, directional, strategic and feasible.

(3) Selection of expertise for strategic research on the S&T roadmap

Scholars in science policy, management, information and documentation, and chief scientists of the middle-aged and the young should be selected to form a special research group. The head of the group should be an outstanding scientist with a strategic vision, strong sense of responsibility and coordinative capability. In order to steer the research direction, chief scientists should be selected as the core members of the group to ensure that the strategic research in priority areas be based on the cutting-edge and frontier research. Information and documentation scholars should be engaged in each research group to guarantee the efficiency and systematization of the research through data collection and analysis. Science policy scholars should focus on the strategic demands and their feasibility.

(4) Organization of regular workshops at different levels

Workshops should be held as a leverage to identify concrete research steps and ensure its smooth progress. Five workshops have been organized consecutively in the following forms:

High-level Workshop on S&T Strategies. Three workshops on S&T strategies have been organized in October, 2007, December, 2007, and June, 2008, respectively, with the participation of research group heads in eighteen priority areas, chief scholars, and relevant top CAS management members. Information has been exchanged, and consensus been reached to ensure research directions. During the workshops, President Yongxiang Lu pinpointed the significance, necessity and possibility of the roadmap research, and commented on the work of each research groups, thus pushing the research forward.

Special workshops. The Editorial Committee invited science policy scholars to the special workshops to discuss the eight basic and strategic systems for China's socio-economic development. Perspectives on China's science-driven modernization to 2050 and characteristics and objectives of the eight systems have been outlined, and twenty-two strategic S&T problems affecting the modernization have been figured out.

Research group workshops. Each research group was further divided into different research teams based on different disciplines. Group discussions, team discussions and cross-team discussions were organized for further research, occasionally with the involvement of related scholars in special topic discussions. Research group workshops have been held some 70 times.

Cross-group workshops. Cross-group and cross-disciplinary workshops were organized, with the initiation by relative research groups and coordination by Bureau of Planning and Strategies, to coordinate the research in relative disciplines.

Professional workshops. These workshops were held to have the suggestions and advices of both domestic and international professionals over the development and strategies in related disciplines.

(5) Establishment of a peer review mechanism for the roadmap research

To ensure the quality of research reports and enhance coordination among different disciplines, a workshop on the peer review of strategic research on the S&T roadmap was organized by CAS Bureau of Planning and Strategy, in November, 2008, bringing together of about 30 peer review experts and 50 research group scholars. The review was made in four different categories, namely, resources and environment, strategic high-technology, bio-science & technology, and basic research. Experts listened to the reports of different research groups, commented on the general structure, what's new and existing problems, and presented their suggestions and advices. The outcomes were put in the written forms and returned to the research groups for further revisions.

(6) Establishment of a sustained mechanism for the roadmap research

To cope with the rapid change of world science and technology and national demands, a roadmap is, by nature, in need of sustained study, and should be revised once in every 3–5 years. Therefore, a panel of science policy scholars should be formed to keep a constant watch on the priority areas and key S&T problems for the nation's long-term benefits and make further study in this regard. And hopefully, more science policy scholars will be trained out of the research process.

The serial reports by CAS have their contents firmly based on China's reality while keeping the future in view. The work is a crystallization of the scholars' wisdom, written in a careful and scrupulous manner. Herewith, our sincere gratitude goes to all the scholars engaged in the research, consultation and review. It is their joint efforts and hard work that help to enable the serial reports to be published for the public within only one year.

To precisely predict the future is extremely challenging. This strategic research covered a wide range of areas and time, and adopted new research approaches. As such, the serial reports may have its deficiency due to the limit in knowledge and assessment. We, therefore, welcome timely advice and enlightening remarks from a much wider circle of scholars around the world.

The publication of the serial reports is a new start instead of the end of the strategic research. With this, we will further our research in this regard, duly release the research results, and have the roadmap revised every five years, in an effort to provide consultations to the state decision-makers in science, and give suggestions to science policy departments, research institutions, enterprises, and universities for their S&T policy-making. Raising the public awareness of science and technology is of great significance for China's modernization.

Writing Group of the General Report

February, 2009

Preface

By 2050, Chinese and global agriculture are expected to enter a new era of development. Global population increase and economic development, particularly in developing countries, will lead to greater human demand for food and fiber. The demand for multifunctional agriculture will also increase as the worsening global energy crisis induces the rise of the biomass energy industry. Demand growth, diversification and market expansion provide unlimited reverie for the future development of agriculture. However, what we should address is that agricultural development will face a series of challenges including global climate change and environmental constraints and competition. With its large population, China will face severe challenges. Therefore, by 2050, the continuing progress of agricultural science and technology will be essential in determining whether Chinese agriculture can successfully seize the opportunity to achieve sustained development for the survival and development of humans.

This is a strong, forward-looking study by the Chinese Academy of Sciences to study and formulate the roadmap for agricultural science and technology development in China to 2050, which is of important strategic significance. The main purpose of the study is to define the challenges and opportunities for global and Chinese agricultural development to 2050, forecast the major demand for agricultural science and technology in the future, make strategic objectives, stage (short, medium and long-term) objectives for agricultural science and technology development to 2050, propose the main direction of agricultural science and technology development in each stage, identify the significant science and technology issues that may achieve breakthroughs, form the overall roadmap for agricultural science and technology development to 2050, and put forward policy recommendations in respect to the institution, resources, and personnel, to achieve these objectives. The report includes:

1) Outlook for agricultural development to 2050 and the demand for technology, general and stage objectives for agricultural science and technology development in China;

2) The overall roadmap;

3) Roadmap for the development of plant germplasm resources and modern breeding technology;

4) Roadmap for the development of animal germplasm and modern breeding;

5) Roadmap for the development of agricultural science and technology based on resource frugality;

6) Roadmap for the development of science and technology of agriculture production and food safety;

7) Roadmap for the development of science and technology of agriculture with modernization and intelligentization;

8) The institutions and policy support for agricultural science and technology development in the future.

In accordance with the overall planning of the roadmap from the Chinese Academy of Sciences, in October 2007, under the common leadership of Qiguo Zhao and Jikun Huang, the Research Group on Agriculture, which consisting of nearly 20 experts from the Chinese Academy of Sciences, began to assume this research task. The research, involving plants, animals, resources, security, modern agriculture, institutional policy, and many other areas is difficult and complex. To this end, the task group adopted such a way of "pool the wisdom of the masses" to work together to carry out this research work, and solicit and learn from outside experts.

Since the task started, the study group has participated in three symposiums organized by Chinese Academy of Sciences in October 2007, December 2007 and June 2008 to define the research intent, significance and research programs of the roadmap for agricultural development. Thereafter, the study group held 2 (in May and November 2008) "Brainstorm" conferences in Nanjing and 3 (in January, March and June 2008) similar conferences in Beijing that invited more than 40 scholars (including academicians, researchers and professors) to give ideas and suggestions. Under the co-authoring of nearly 20 experts in many areas, after repeated arguments and amendments, the first draft of the roadmap for agriculture was finally completed in November 2008 and approved in the review conference for the roadmap for agriculture from the Chinese Academy of Sciences in Beijing in that same month. The chairs of the review conference were academicians Zhensheng Li, Zhizhen Wang, Shengli Yang, Xiao Li, Yang Sheng and Zhibin Zhang. According to the advice of the experts, the group held a seminar for the draft amendments in Nanjing in February 2009 and identified the unification plan. After the final draft was completed, the study group invited the five academicians and experts Honglie Sun, Taolin Chang, Zhihong Xu, Zuoyan Zhu and Yiyu Chen for peer review and made the last modifications and improvements according to their valuable comments.

With the members of the Research Group on Agriculture helping each other and working together, the research work of the roadmap for agricultural science and technology development lasted for one and a half years. Events that have happened forty or fifty years ago are often difficult to understand and have to be evaluated carefully by historians from time to time. The roadmap for agricultural science and technology development is perhaps even more difficult to accurately report as it depicts the blueprint for the development of agricultural science and technology in forty or fifty years. Dozens of domestic experts in related fields devoted their wisdom and energy to this report. However, in face of the unpredicted future, human cognition is very limited. No one is of exception. Scientists are not prophets, but there is one thing we have been able to confirm, and that is the reality will certainly not entirely be realized as we wrote in the book. Therefore, if you are inspired or have learned anything after reading this book, the efforts all the members of the group have made will be worthwhile!

Research Group on Agriculture of the
Chinese Academy of Sciences

December, 2010

Contents

Abstract

In the next 50 years, world agriculture will be gradually entering a new epoch. Economic growth will substantially contribute to the global demand for food and fiber. Food consumption pattern will change significantly, and the demand for agricultural multifunction will also increase as well. The intensification of global energy crisis will induce the rise of agricultural bio-energy industry. Agriculture will not only continue to play the traditional roles in ensuring food security and national economic development, but also take up new historical missions to mitigate global energy crisis and provide a favorable environment for human survival. Agriculture will be expanded with substantial demand for its products and service in the future. However, development opportunities and great challenges always coexist. By 2050, agriculture will have to meet the demand from 9 billion people for food and fiber in quantity and quality on the one hand, and on the other hand, the rise of bio-energy will be a direct threat to the food security; meanwhile, agriculture will face various of risks brought about by global climate change and environmental degradation of resources. Therefore, in the next 50 years, science and technology will play a crucial role for agriculture to meet the demand from human society and economic development.

In the next 50 years, along with the development of global agriculture, agriculture in China will also be gradually entering into a new era. On the demand side, total demand for food and fiber will increase significantly, except for the demand growth for rice and wheat that will show a slow decline trend in the coming ten years. The demand for dairy products and aquatic products is project to rise more than 3 times, and the demand for livestock, feed grain, fruits, edible oil and fiber aggregate will increase by 1.5-1.6 times, vegetables and sugar will increase by 75% and 1-fold respectively. In addition, the demand structure for agricultural products will also experience a fundamental change in China in the next 50 years. The main driving forces for the demand increase and the structural change are income growth, urbanization and population growth. By 2020, under the pressure of resource shortage, demand increase and demand structure changes, agriculture in China will continue to play its traditional role to ensure the national food security, improve people's nutrition and maintenance the livelihood of farmers, meanwhile, agriculture will begin to undertake to improve the eco-system, mitigate climate change, provide tourism agriculture and rural landscape and inherit the cultural and traditional knowledge. After 2030, the new functions of agricultural diversification will continue to strengthen.

In such a trend, agricultural development in China will face a great deal of chances and challenges in the next 50 years. The facts that the expansion of agricultural markets, increased farming efficiency, the improvement of agricultural production structure and the development of agricultural science and technology, etc. will provide a rare opportunity for agricultural development in the future; meanwhile, a series of challenges will arrive successively, which includes the enormous challenges of agriculture to meet the increasingly increased domestic demand for agricultural products in quantity, quality and safety; agricultural development is facing increasingly severe challenges of land and water resources and ecological security; agricultural development is facing the increasingly severe challenges of the conflict between small-scale operation and agriculture modernization; threats and challenges to food security and overload pressure over water and soil resources brought about by the expansion of global and domestic demand for agriculture-based bio-energy and agricultural multifunction in China; the impact and influence on agriculture brought about by global climate change. Obviously, the development of agriculture in China in the future needs to make a breakthrough in the fields of production increasing potentiality of plants and animals, resource management utility and security, agricultural production and food safety, ecological system and environmental protection to meet the rapid social and economic development. To this end, it is necessary to make a roadmap for agricultural science and technology development to effectively promote the development of agriculture in China.

According to the technical demands mentioned above, this report made a roadmap for agricultural science and technology development in China to 2050. This roadmap mainly includes the following five areas of sciences and technologies: plant germplasm resources and modern breeding; animal germplasm and modern breeding; resource saving; agriculture production and food safety; and agricultural modernization and intelligentization. The overall goal of this roadmap is as follows: by 2050, with the great innovation and breakthroughs in the above agricultural science and technology areas, agriculture in China will possess the technological supporting to realize the sustainable use of agricultural resources, ensuring and improving the national food security, and successfully transferring into the new era of the coexistence of traditional function and modern multi-function in the future.

To meet this overall goal, the five major areas of agricultural science and technology have developed their own overall goal and roadmap to 2050 respectively:

1) In the field of plant germplasm resources and modern plant breeding science and technology, using the methods of system biology mainly and the products-oriented strategy, exploring the key and functional genes in germplasm and utilizing the superiority of gene resources-rich in China, making breakthrough in plant photosynthesis research and developing genomic-

knowledge based key biotechnology, constructing the new innovation system of functional plant products development, providing the scientific supporting to agricultural sustainable development.

2) In the field of animal germplasm resource and modern breeding science and technology, we mainly use integration of multidisciplinary research methods in the fields of life science and biotechnology such as systems biology, bioinformatics, genomics and proteomics, genetic engineering, etc, with major product-oriented research & development strategy, and develop healthy and sustainable development of animal aquaculture, including marine fisheries, in order to cultivate livestock, poultry and aquatic products with security, quick growth, high protein-content, high meatyield, high feed transformation or resistance to diseases.

3) In the field of resource saving agricultural science and technology, it is to develop high-efficient utilization mechanisms and methods on water, nutrients, and artificial assistance resources. Main investigations focus on arable land nourishing and substitution technology, engineering-biology-chemistry water saving technology, watershed water resource management technology, water-fertility-energy integrated management, precision mechanized implement technology, and intelligent fertilizer. It is to establish three agriculture production technology systems, e.g., land-saving agriculture, water-saving agriculture, fertilizer- and energy-saving agriculture. It is to realize agriculture production intensification, mechanization, large-scale operation and industrialization. It is to realize the enhancement of sustainable agriculture production and the utilization of resources.

4) In the field of agricultural production and food safety science and technology, it is going to enhance the basic theoretical researches on agricultural production safety, prevention and control of major pests and diseases; maintain the nutrition of agricultural products, clean control, storage and processing and so on, the key technological breakthroughs and integrated technologies integrative innovation, the establishment of the prevention and control warning system of agricultural pests and diseases, intelligent expert management system, and form the food safety digital tracing system from farm to fork, implementation of accurate monitoring and prevention and control of "active security strategy". to establish a standardized system of safe production of agricultural products and green environment, to ensure the quality and safety of agricultural products in the cultivation, breeding, storage and processing; to achieve precise design and quality regulation of food, and proceed the various nutrition food R&D to improve the diet combination; create "intelligent personalized nutritional food" to meet the personalized nutritional needs and provide the functional foods for the purpose of enhancing the physical quality and fitness of the whole population.

5) In the field of agricultural modernization and intelligentization science and technology, it is to establish a modernization platform for promote innovations on Chinese agricultural researches and realize the following goals

through the breakthrough of the key technologies and equipped by high-tech. These goals are: agricultural information service network, digital management of agricultural resources, precision management of the agricultural production process, intelligent of agricultural equipment and agricultural machine, network platform for virtual research in agriculture, rapidly increasing agricultural productivity, resources efficiency and agricultural continuous innovation ability.

In order to ensure the achievement for overall goal of agricultural science and technology development and according to the requirement of roadmap design plan, this report separates the general agricultural technology development and sub-areas technology development into certain objectives for phases by short-term (2020), mid-term (2030) and long-term (2050).

The goal of agricultural science and technology development to 2020 mainly are: establishing the gene bank and database for ecological populations and germplasms and its information sharing-system, developing multigene transgenic and multitraits improved cultivars, perfecting the technology of genetic transformation and elite germplasm innovation in main crops, combing with molecular marker assisted selection and safe transformation technology, realizing the high-efficiency transformation and congregation of major-effect gene and its interacting network; establishing important livestock, poultry and aquatic animals germplasm resource sharing platform, develop molecular marker technology and explore special value of germplasm resource and enhance genetic improvement strength and potential innovation of the fine varieties of livestock, poultry and aquatic animals; developing primarily three agriculture production technology systems, e.g., land-saving agriculture, water-saving agriculture, fertilizer and energy-saving agriculture in order to establish the production base for sustainable agriculture development; establishing the standardized system of agricultural production safety, distribution & preserve and processing, and it is not only structuring the green environment for green food production, but also providing green safety agricultural products with high-quality; completing the development of multi-functional agricultural information network platform and professional search engine, establishing a large regional scale professional database and agricultural information service network, proceeding the researches on key technologies of digital resource management and precision management of the agricultural production process, intelligent of agricultural equipment and prototype of the network platform for agricultural virtual.

The goal of agricultural science and technology development to 2030 mainly are: improving the technology of molecular design, hasting the technology of plant breeding by molecular design, developing and releasing new cultivars for crop production; obtaining primarily the new crops by design and assembling, breeding and releasing new super-energy plant cultivars; establishing the molecular design of livestock, poultry and aquatic animals breeding technology with Chinese characteristics and breed new varieties of pig, cattle, sheep, chicken, fish, shrimp, shellfish and other key breeding animals with quick growth, high protein content, high meatyield, high feed transformation or resistance to diseases; perfecting three agriculture production technology systems, e.g., land-saving agriculture, water-saving agriculture, fertilizer and energy-saving agriculture in order to realize leapfrog development of intensification, mechanization, standardization and industrialization for agriculture production; realizing the precision control of storage, logistics and processing quality of agricultural products, establishing the management system of intelligent environmental monitoring and remediation of contaminated soil, creating a sustainable ecological environment, developing processed foods with various nutrition and provide a variety of nutrients that human health needed; realizing the agriculture information service network around China, realizing quantitative management of soil, water and weather resources, realizing dynamic monitoring of farmland water, pest and weed, realizing intelligent precise management of farmland water, pesticide, cultivation and breeding, and establishing the network platform for agricultural virtual research.

The goal of agricultural science and technology development to 2050 mainly are: realizing genome-wide gene optimization and assembling, developing smart new plant cultivars with high yield potential, good end-using quality and multiple functions, developing and releasing a large number of high-energy content plant cultivars for industry of bio-energy, developing effective ecology strategies to control livestock, poultry and aquatic animals disease, and develop suitable vaccines and effective drugs; discovering a crop of effective drugs with high value of output and no harm to human health, and then achieve the ecological management of aquaculture; establishing overall three agriculture production technology systems, e.g., land-saving agriculture, water-saving agriculture, fertilizer and energy-saving agriculture in order to stabilize the sustainable agriculture development; developing a harmonious agricultural environment which can satisfy multi-needs, setting up the design standards of personalized nutritional foods, creating “intelligent personalized nutritional food” to meet the individual nutritional needs; completing the construction of varies scale information collection system and realize the agricultural resources digital management basically, conducting the precise management to animals and plants during the production process, realizing comprehensively agricultural informationization and precise management, and realizing the agricultural research innovation through the network platform of virtual agriculture.

Series of institutional and policy supports are needed to realize the roadmap for the agricultural science and technology development to 2050. To guarantee the achievement of the roadmap for the development of agricultural science and technology to 2050, it is urgent to deepen reform the national agricultural science and technology system, clearly identify the roadmap of agricultural science and technology innovation toward 2050 and in each sub-period, establish an effective incentive mechanism to improve the initiative of scientific and technical personnel, and create a better national agricultural science and technology innovation system; it is also urgent for the country to establish a long-term investment mechanism to increase investment on agricultural science and technology, and improve its investment structure by raising the ratio of core funding in the total investment. In addition, it is also urgent to improve relevant laws and regulations (e.g., intellectual property right protection, environment legislation, etc.) and establish national agricultural science and technology talent funds to attract a batch of excellent leading talents and form innovative teams to enhance the independent innovation ability of agricultural science and technology in China.

1 Outlook of Agricultural Development to 2050 and the Demand for Science and Technology

1.1 Outlook for Global Agriculture Development to 2050

1.1.1 Human Demand for Food and Fiber Will Continue to Expand, and the Consumption Structure Will Change Significantly

By 2050, the world demand for agricultural products such as food and fiber will increase significantly. The rapid growth in demand for agricultural products will mainly occur in developing countries. The demand for food and fiber has stabilized and grown slowly in developed countries, but continuous population expansion, economic growth and urbanization in developing countries will increase the demand for food and fiber in quantity and quality while changing the global food consumption structure.

1. The expansion of world population will continue to increase the demand for food and fiber, and nearly 90% of the increase in population will be in developing countries

According to UN[1], the world population will increase by 40% from 6.4 billion in 2005 to 9.0 billion in 2050. Population expansion will continue to increase the demand for food and fiber. From a geographic perspective, of the additional 2.6 billion people, developing countries account for more than 2.3 billion. Therefore, the global growth in demand for agricultural products caused by population explosion will mainly be due to the population growth in developing countries.

2. Economic growth and urbanization in developing countries will further and significantly increase the per capita demand for food and fiber

International Food Policy Research Institute predicts that from 2000 to

2050, global food demand will increase by 75%, and the demand for meat will double. In terms of food consumption, the demand for high-valued agricultural products (such as livestock, aquatic products, vegetables and fruits, etc.) will rise with the growth of incomes. International Food Policy Research Institute recently projected that by 2050, per capita consumption of livestock products in South Asia, Sub-Saharan Africa, Central Asia and North Africa, East Asia and the Pacific region will be doubled in 2000-2050 (Fig. 1.1)[2].

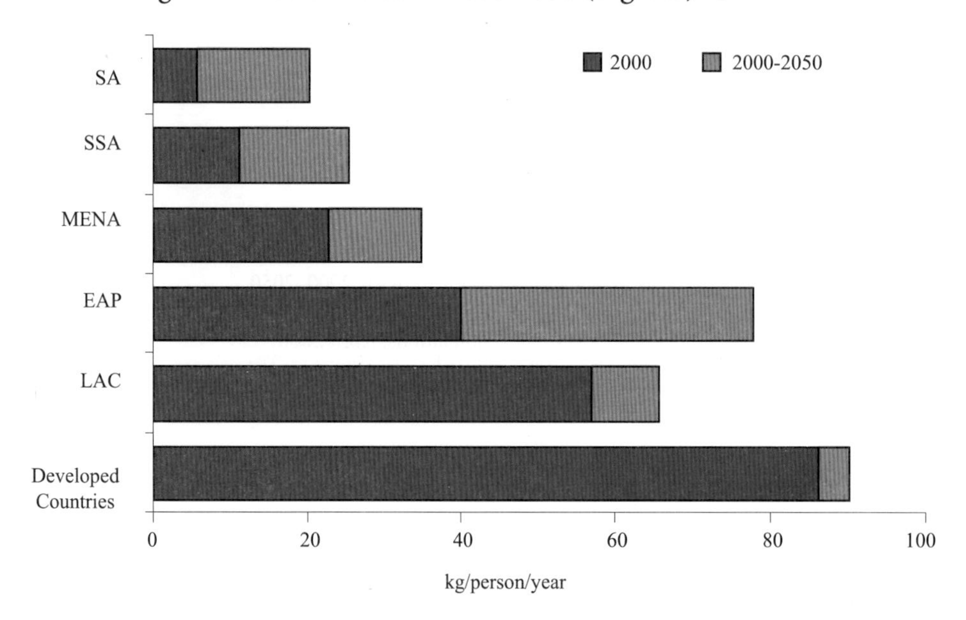

Fig. 1.1 The consumption of animal products per capita from 2000 to 2050
SA. South Asia; SSA. Africa south of the Sahara; MENA. Middle East and North Africa; EAP. East Asia and Pacific; LAC. Latin American and Caribbean.
Source: IFPRI IMPACT simulations, April 2008

3. Economic growth and accelerated urbanization will raise the demand for improved quality and safety of food and food consumption structure will further change

With incomes increasing and the acceleration of urbanization, the demand for food quality and food safety from consumers will continually rise. In addition, the consumption level of high-valued agricultural products will also continue to increase, and food consumption structures will change significantly. The consumption of livestock, aquatic products, vegetables and fruit will grow and will replace part of staple food consumption; the per capita cereal food consumption (or direct food consumption) will only increase slowly. To 2050, per capita cereal food consumption will even present a downward trend in Asia (Fig. 1.2)[2].

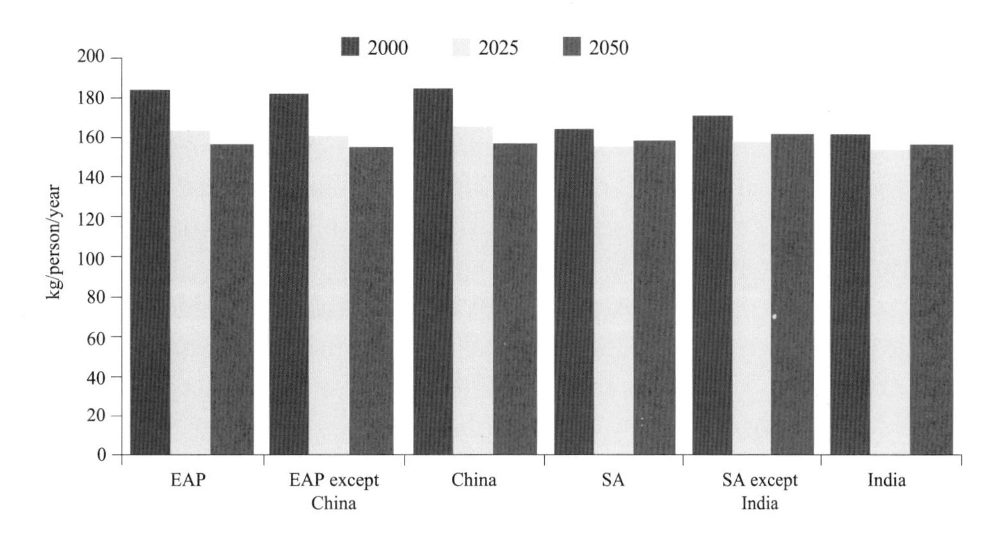

Fig. 1.2 Per capita food demand, Asia, 2000–2050
EAP. East Asia and Pacific; SA. South Asia.
Source: IFPRI IMPACT simulations, April 2008

However, although the per capita demand for rice and wheat in many Asian countries will gradually decline, the increase in meat consumption will bring a substantial increase in demand for maize and other feed grains. Coupled with impact of population growth, the quantity of the demand for grain (food and feed) will continue to expand (Fig. 1.3)[2].

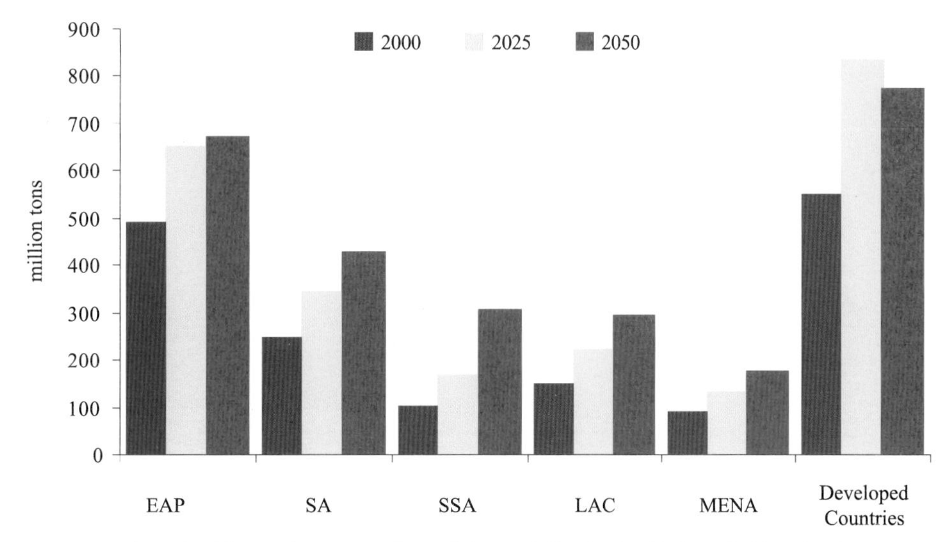

Fig. 1.3 The total amount of cereal demand (food+ feeds) from 2000 to 2050
EAP. East Asia and Pacific; SA. South Asia; SSA. Africa south of the Sahara;
LAC. Latin American and Caribbean; MENA. Middle East and North Africa.
Source: IFPRI IMPACT simulations, April 2008

1.1.2 Demand for Multifunctional Agriculture (Such as Bio-energy, Good Habitat, Landscape, Etc.) Will Continue to Expand

By 2050, humans will have a higher demand for both production and living environments, which will further exacerbate the global energy crisis. Energy prices are likely to maintain an upward trend in the long term. Agriculture will not only continue to play its traditional roles to protect food security and the basis of a national economy, but agriculture will also take over other new historical missions of providing renewable resources to relieve the energy crisis and better the environment for human livelihood. Then, global agriculture will enter a new era.

The energy crisis induces the rise of the agricultural bio-energy industry. Since the late 20th century, prices for global energy have been continually rising, which makes many of the energy importers fear for their energy insecurity. Concerns over energy insecurity also contribute to the rise and rapid development of the modern bio-energy industry [3-5]. The debate over the development of biofuels, which uses agriculture, forestry and other biomass as raw materials, has heated up recently. However, driven by strong political, energy security and economic interests, biofuels will become a dynamic new sector [5, 6] in the future of national economies. The United States and Brazil are the largest biofuel (ethanol) producing countries with a total output of 850 million and 500 million gallons of biofuel (ethanol) in 2007, respectively, By 2020 these countries plan to increase production to 2.2 billion gallons and 1.3 billion gallons (Fig.1.4)[7-10] respectively. The development of the agricultural bio-energy industry is bound to lift the demand for (put forward a higher requirement on) agricultural production, and opportunities and challenges coexist.

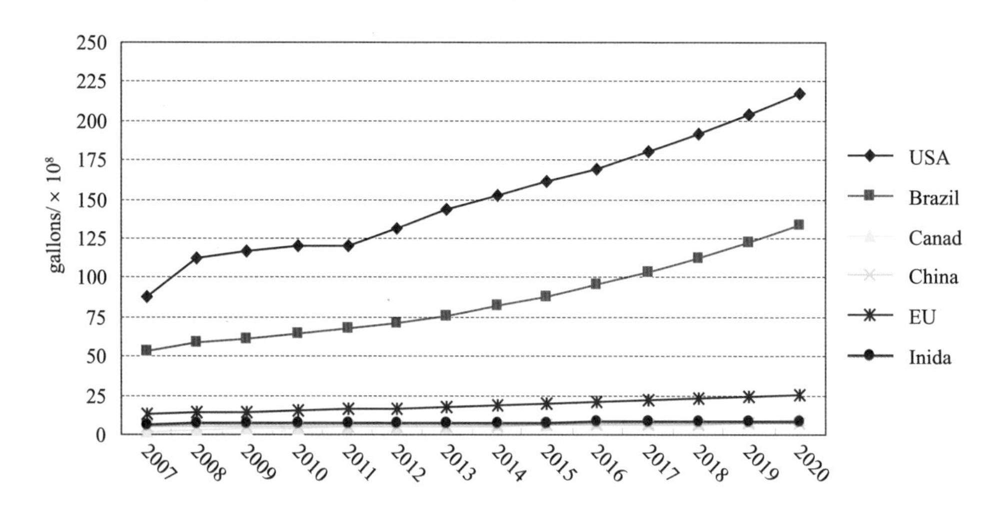

Fig. 1.4 Current and projected total bioethanol production targets in major countries

Economic development induces demand from human society for multi-functional agriculture[11]. The demand for multi-functional agriculture includes improving food security and environmental, cultural, and traditional knowledge preservation. In terms of the environment, multi-functional demand includes ecological system improvement, climate change mitigation (carbon sequestration, land rehabilitation), biodiversity maintenance, water and soil conservation, control of air quality (such as the reduction of greenhouse gases), and ecotourism. In terms of agricultural heritage in regards to cultural and traditional knowledge, the multi-functional demand includes agriculture diversifying the lifestyle of the community, maintaining social stability, and protecting personal values, families and traditional lifestyles in the community.

1.1.3 The Development Opportunities for Global Agriculture to 2050

By 2050, global agriculture will face a series of development opportunities, of which the following five aspects are particularly prominent.

1. The increasing demand for food and fiber from humans provides space for agricultural production and agricultural market expansion

The development of any industrial sector is closely related with social and market demand for its products and services. With rise in population growth and income, by 2050 the global demand for food and fiber will increase significantly, which will expand demand markets for many major agricultural producers in the world.

2. Agriculture is likely to become an industry with substantial developing space with the increasing demand from humans for multifunctional agriculture

The increasing demand from humans for agricultural products, good rural landscape and better living environment, and bio-energy is likely to lift agriculture into a new era of development. The role of agriculture within social development and a national economy will gradually extend from ensuring food security to also preserving positive environmental externalities, traditional knowledge and cultural heritage.

3. Agriculture is likely no longer be a low comparative advantage industry, its relative comparative advantage will increase significantly

The declining trend in the prices of agricultural products over the past 100 years will be reversed, and prices of agricultural products will keep rising despite rising frequency of fluctuations[10,11]; the rise in the prices for agricultural products will help to improve the comparative advantage of agricultural production, and agriculture and related industries are likely to attract a lot of public and private investment to enhance agricultural productivity and the relative comparative advantage of agriculture.

4. Economic globalization will improve the efficiency of agricultural resource uses and global agricultural trade, and promote global agricultural development

Economic globalization will be one of the main themes of economic development in the 21st century, which challenges many countries' agriculture and also supplies vast opportunities for it. By 2050, the agricultural development in various countries and regions will be largely based on resources and the comparative advantage of agricultural products. International trade will play an important role in national agricultural production in various countries.

5. To 2050, many areas of agricultural science and technology have great potential to develop

Should the state and enterprises continue to increase investment, agricultural science and technology will inevitably become the most important driven force to increase agricultural productivity. We project that, by 2050, the new technologies rapidly rising in molecular biology, genetic engineering and natural resource management in company with traditional breeding techniques will become the main technologies to improve agricultural productivity in the coming decades[11,12].

1.1.4 The Main Challenges in the World to 2050

By 2050, global agriculture will face many development opportunities, but it will also face a series of huge challenges. Not only do the opportunities and the challenge coexist, as mentioned previously, but also agriculture will face many challenges and risks brought about by climate change and resource and environmental degradation.

1. It will be a great challenge to meet the demand for food and fiber in both quality and quantity from 9 billion people by 2050

With the world population increase, economic growth and urbanization, the demand for food will double, which will bring about increasing pressure on water resources and arable land. Marginal cost of investment in R&D to raise output in per unit land is expected to rise. Meanwhile, the increasing demand for feed, due to the rapid growth of demand for animal products, will make countries increasingly pay more attention to the coordination of the supply structure of food and feed grains.

2. The development of an agricultural biofuel industry provides development opportunities for agriculture, and also presents a series of challenges

Under global energy crisis, the competition between the demand for agricultural products from the agricultural biofuel industry and the demand for food will become increasingly fierce, and food security will face unprecedented challenges in developing countries. From 2006 to 2007 the expansion of an agriculture-based biofuels industry brought about the universal increase in food prices. According to IMF, in 2006 and 2007, global food prices rose 10.5% and 21.6%, respectively. The most fundamental reasons for the rapid rise in food prices during this period are the soaring energy prices and the biofuel expansion. By 2020, the United States is expected to use about 150 million tons of maize to produce fuel ethanol, accounting for around 38% of U.S. domestic production. How to deal with the problem of biofuel expansion competing for food with humans, for land with grain production, and feeds with animal production will be the new and great challenges for agricultural development to 2050.

3. Water shortages, depletion of non-renewable energy resources, land and forest resource degradation threaten the sustainable development of agriculture

In the past, many countries increased agricultural production largely at the expense of the environment and non-renewable energy sources (Fig. 1.5)[2], and this trend is likely to continue in the future. Resource degradation and depletion will lead humans to develop new alternative energy; meanwhile, the new combination of resource utilization will present new challenges for the sustainable use of resources.

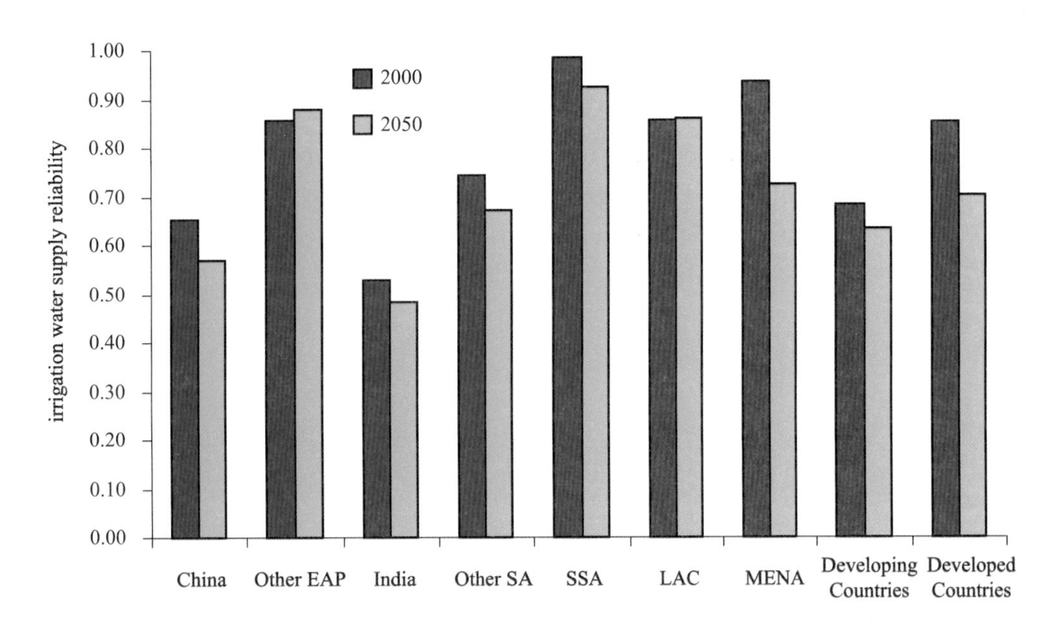

Fig. 1.5 The change of irrigation water supply reliability from 2000 to 2050

EAP. East Asia and Pacific; SA. South Asia; SSA. Africa south of the Sahara; LAC. Latin American and Caribbean; MENA. Middle East and North Africa.

Irrigation water supply reliability is defined as the ratio of actual irrigation water consumption to potential irrigation water consumption.

Source: IFPRI IMPACT simulations, April 2008

4. Climate change and the ecological environment pose new challenges to agricultural production

IPCC reports that the level of food imports in many developing countries will keep on increasing in the future. Global agricultural GDP will decline 16% in the next 20 years due to global warming. Although the impacts of climate change differ largely among regions (either positive or negative impacts), more than 40 developing countries will fall in food production; production fall in some countries in South Asia will be great, some reaching 22%[13]. Of course, in some areas, climate change may bring benefits (such as developed countries and South America). Geographically, semi-arid regions like Africa may increase planted area. However, the increase in cultivated area may not be utilized under the premise of diminishing water resources, which will bring about new challenges and complexity to the development of agricultural science and technology. Climate change will also leave the ecological environment increasingly vulnerable, and the frequency of extreme weather and disasters will increase rapidly, which will impact agriculture and also demand responses, including improved accuracy of forecasts, research and response strategies.

5. By 2050, increases in agricultural production will mainly depend on productivity improvement, and whether or not technological progress can undertake this difficult mission will be a major challenge to science and technology community

International Food Policy Research Institute (IFPRI) projects, to 2050, except for Sub-Saharan Africa, Latin America, the Caribbean region, Central Asia and the North Africa region where crop area will continue to expand, the area of crops in all other regions will keep a long-term downward trend. In all the countries of the world, the increase of cereal production will mainly depend on the improvement of unit yield (Fig. 1.6)[2], while the unit yield increase to a large extent requires strong technological support. Therefore, whether technological progress can assume this arduous task will be the major challenge to global agricultural development in the future.

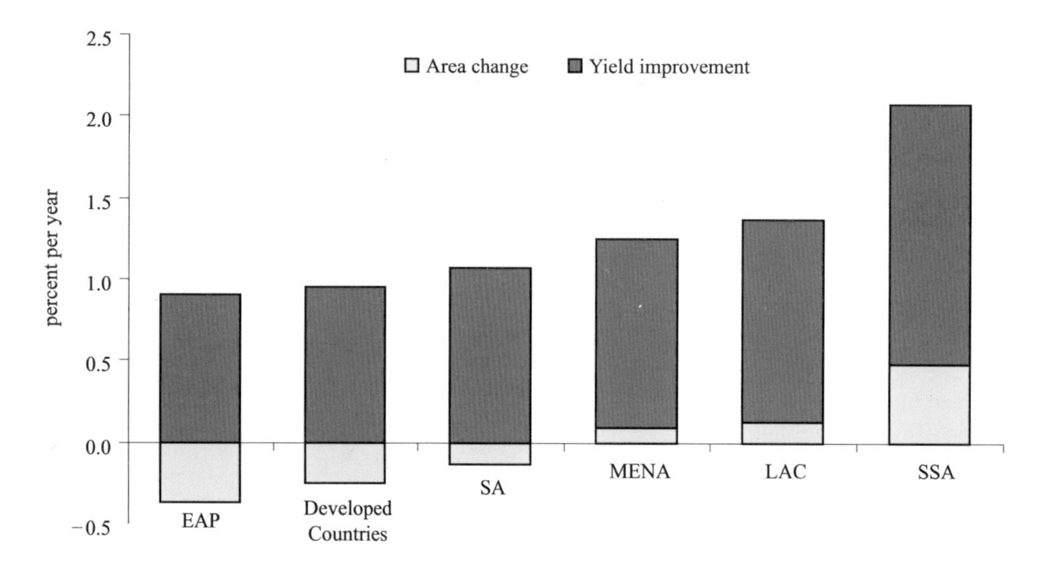

Fig. 1.6 Sources of cereal production growth, baseline, 2000-2050

EAP. East Asia and Pacific; SA. South Asia; SA. South Asia; MENA. Middle East and North Africa; LAC. Latin American and Caribbean; SSA. Africa south of the Sahara

1.2 Outlook of Agricultural Development in China to 2050

Over the past 50 years, China's agricultural growth has been impressive and has attracted the world's attention, but China has also paid a high environmental price for it rapid growth. Over the past 50 years, China has made significant efforts to supply appropriate and adequate food and clothing for its growing population. By the early 21st century, China supported 1/5 of the world population with only 1/15 of cultivated land in the world, and the overall food self-sufficiency rate was over 100%. The total number of undernourished people decreased from 193 million in the early 1990s to less than 100 million in the early 21st century. Over the past 50 years, China has made great achievements in agricultural development but also paid a high environmental cost. The resource base for agricultural survival is deteriorating. In the next 50 years, China will also face tremendous opportunities and great challenges in agricultural modernization and sustainable development.

1.2.1 The Overall Demand for Food and Fiber Will Increase Significantly in China, and Food Consumption Structure Will Change Fundamentally

By 2050, the demand for dairy products will increase by more than six times in China; the demand for seafood will increase by nearly 3 times; the total volume of demand for livestock, feed grains, fruits, cooking oil, and fiber will increase by 1.5 to 1.6 times, the demand for vegetables and sugar will increase by 75% and 1 fold respectively, and the demand for rice and wheat will slowly increase before presenting a declining trend. In addition, by 2050, the structure of demand for agricultural products in China will undergo fundamental changes. The main driving forces for the changes of the total volume of the demand for food and the demand structure are income growth, urbanization, and population growth.

1. The demand for livestock, aquatic products and feed grain will increase significantly, and income growth and urbanization will be the main driving forces for the increase in per capita demand for animal products, aquatic products and feed grain

Center for Chinese Agricultural Policy projected that per capita annual demand for meat and meats in urban and rural will increase from 31 kg and 54 kg in 2004 to 69 kg and 89 kg, respectively, in 2050 (Table 1-1). As urban residents and rural residents keep different consumption patterns, accelerated urbanization processes will further increase the demand for meats and aquatic products nationwide. Income growth together with urbanization will increase the per capita demand for meats from 41kg in 2004 to 84 kg in 2050 nationwide (Table 1-1). Among animal products, the demand for dairy and aquatic products increase fastest, and by 2050, the per capita demand will increase to 100 kg and 48 kg, respectively, both from the about 15-16 kg in 2004 (Table 1-1).

Table 1-1 Per Capita Annual Food Consumption in Urban and Rural China, 2004 to 2050 (Unit: kg/person)

	Rural				Urban				The National Average			
	2004	2020	2030	2050	2004	2020	2030	2050	2004	2020	2030	2050
Rice	96	92	88	85	50	45	39	34	76	66	57	47
Wheat	85	82	79	77	39	36	32	29	65	57	50	41
Edible oil	8	13	15	17	14	21	22	23	11	17	19	21
Vegetable	158	189	192	196	187	229	242	259	171	211	223	243
Fruit	23	35	41	46	75	104	109	120	46	73	83	101
Poultry	31	48	54	69	54	72	80	89	41	61	70	84
Milk	3	20	40	72	33	75	91	110	16	50	72	100
Aquatic Products	9	17	21	29	23	38	44	55	15	29	35	48

Source: Huang J K. China's Agriculture toward 2050. A report submitted to IAASTD.Center for Chinese Agricultural Policy, Chinese Academy of Sciences. Beijing. 2008.

Population in China will shift from a slow and declining growth to a slight decline and negative growth in 2000-2050. The increase in demand for agricultural products due to population change will mainly occur before 2030. United Nations and the Chinese government projected recently [1] that the population in China will reach a peak at 1.45 billion by 2030 and fall down thereafter. But the huge population base will impose enormous pressure upon food demand in China for a long period. Considering the impact on per capita demand for livestock products and aquatic products brought about by income growth and urbanization combined with the impact on gross demand brought by population growth, by 2050, the demand for aquatic products will increase by nearly 3-folds, and animal products and feed grain will increase by 1.5-1.6 times.

2. The demand for fruits, vegetables, cooking oil and sugar and other food will keep increasing, and the production structure of crops will undergo a fundamental change in China in the coming decades

By 2050, the main driving forces for the increase in per capita demand for fruits, vegetables, edible oil and sugar will also be income growth and urbanization. According to our projection, by 2050, per capita demand for fruit and edible oil will increase by one fold or more, and vegetable and sugar will increase by around 50% (Table 1-1). Population growth will continue to raise the demand for these products. By 2050, the total volume of the demand for fruit and oil will increase by nearly two fold in China, and sugar will increase by around 1 fold, vegetables around 75%.

But the concern is that income growth and urbanization have not brought about the increase of per capita demand for rice and wheat, major staple foods in China. We projected that per capita consumption of rice and wheat will continue on a downward trend that has characterized the past 10 years mainly due to the increased consumption of animal products and other high value-added agricultural products due to income growth and urbanization (Table 1-1). However, the continued growth of the population will also allow for the total rice and wheat demands to keep on slowly increasing until 2020.

3. Economic growth and the development of garment and textile industries will greatly increase the demand for cotton and other fiber crops in the market

By 2050, the demand for cotton and other fiber products in China will undergo the impact from increased domestic demand due to improved living standards, and with the accelerated process of trade liberalization, the expansion of export-oriented apparel and textile industry in China. The development of garment and textile industries will greatly increase the demand for fiber crop products in the market. According to our projection, even if the domestic production of cotton and other fiber crops products will be doubled by 2050, which can only meet 60% of the total demand (including domestic demand and export demand), cotton imports will gradually increase, and cotton will become one of the major imported agricultural products in China in the future.

1.2.2 The Sustained and Rapid Growth of China's Economy Will Gradually Increase the Demand from the State and Society for Agricultural Multifunction

By 2050, China's economy will gradually stride into the ranks of developed countries from the current middle-income developing economies and in this long-term and sustained process of economic development, the national and social demand for agricultural multifunction will gradually increase. The sustainable growth of Chinese economy is also closely related to the national energy security; meanwhile, the rapid development of the Chinese economy and the improvement of people's living standard will increase the demand from the whole society for the improvement of agricultural production and living environments.

We projected that by 2020, under a series of pressures of limited resources, increased demand and diversification, China's agriculture will continue to play its traditional role to improve food security and human nutrition, and maintain the livelihood of farmers. Agriculture will also begin to improve the ecological system, mitigate climate change, provide agricultural tourism and rural landscape and provide cultural and traditional knowledge heritage. After 2030, the demand for agricultural multifunction will gradually increase, and agriculture will further undertake new historic missions to provide renewable energy to mitigate the energy security and provide a beautiful living environment for the citizens. By 2050, China's agriculture together with the rest of world will also enter into a new era of agricultural development, that is agriculture will not only continue to play its traditional role to improve food security and national economic development, but it will also play an active and important role in solving the global energy crisis and providing a beautiful environment for human survival.

1.2.3 Major Development Opportunities for China's Agriculture

By 2050, similar to global agriculture, agriculture in China will also face many development opportunities. The market expansion, increased comparative advantage of agricultural sector, improved agricultural production structure and technological development will especially provide a rare opportunity for agricultural development in China in the future.

1. China will share the opportunities of agricultural expansion due to the increased demand for food, fiber and agricultural multifunction with many other countries in the future

In terms of the increased demand for food and fiber in the future, China will be one of the countries whose demand for agricultural products such as livestock products, aquatic products, fruits, edible oil, sugar and cotton will have the fastest growth rate,and the highest total increased demand in the world. The demand for agricultural multifunction has just started, and in the future, the increasing demand for agricultural multifunction will greatly expand space for agricultural development.

2. The structure of agricultural production in China will be continuously improved, and economic globalization will provide greater potential space for the China's agricultural products with relatively comparative advantage

The agricultural products with comparative advantage in China include animal products, vegetables, fruit and flower industries. Economic globalization will play a positive role in promoting production in line with their comparative advantage. At the same time, as the prices of agricultural products in the international market present a long-term upward tendency, agriculture will likely attract more public and private investment in agricultural and rural infrastructure, which will facilitate agricultural productivity growth.

3. By 2050, the capacity of agricultural science and technology innovation will be improved greatly in China, and the overall level of agricultural science and technology in China will reach the world's leading level

With the implementation of medium and long-term planning for technological development nationwide, China's R&D investment will keep rising. And with the establishment and improvement of national agricultural science and technology, the process of agricultural modernization in China will further accelerate and the status of agricultural science and technology as the engine of agricultural productivity growth will be further raised. Although the land productivity of rice, wheat, maize, cotton, oil and sugar crops and other major crops in China has reached a relative high level, the significant growth crop yield has become more and more difficult. However, many studies have shown that there is great potential for Chinese crops to increase the yield and to improve the efficiency of agricultural resources.

1.2.4 To 2050, China's Agriculture Will Also Move toward an Ecological High-value and High-tech Agricultural System

Ecological high-value agriculture, a high-tech agricultural system and production mode fully equipped with modern and future new energies, materials, facilities, IT and biological technologies, aims at achieving high-value agricultural industry by improving the technological contents of agriculture and the level of agricultural management under the premise of ensuring a good eco-environment so as to effectively improve the productivity, and the level of being industrialized, competitiveness and comparative effectiveness of agriculture. The key mission of the ecological high-value agriculture is to achieve the ecological and high-value utilization of agricultural biology, water and soil resources and wastes, to obtain the ecological and high-value agricultural production, and to realize the ecological and high-value processing and marketization of agricultural products after the production. Industrial systems of the ecological high-value agriculture mainly include the following four aspects, namely of agricultural products safety, sustainable agriculture, smart agriculture and high-value agriculture. Ecological high-value agriculture is a general concept that integrates ecological agriculture and environment, agricultural products being high-yield, high-quality and high-effectiveness with integrated technological, market-oriented and industrial economic values. It is the development direction of modern agriculture.

1.2.5 Major Challenges in Agricultural Development in China

By 2050, in the face of development opportunities, Chinese agriculture will also face a series of challenges. The challenges for Chinese agriculture will be more difficult than those for many other countries. By 2050, besides the five major challenges in global agricultural development, Chinese agriculture will have to face the following challenges related closely with the national conditions of China.

1. Chinese agriculture will face enormous challenges to meet growing domestic demand for agricultural products in both quantity and quality and to improve its agricultural security

China has the largest population in the world. In despite of the fact that the population growth rate will slow down in coming years and that the total population will reach the peak before 2050, the huge population base will pose tremendous pressure upon food demand in China in the long term. By 2050, Chinese agricultural will not only have to play the traditional role to increase production and improve human nutrition, but it will also have to adapt to the structural changes of food consumption brought about by improving living standards and the demand for food safety and agricultural multifunction.

If technological development and agricultural inputs still stay at the current level by 2050, with the increase of domestic demand for agricultural products, the self-sufficiency rate of many domestic products without comparative advantage will decrease significantly, and the supply of these agricultural products would be subject to a certain degree of security threats. According to Center for Chinese Agricultural Policy in the Chinese Academy of Sciences, under business as usual (or under the baseline scenario, i.e,. by 2050, technological development and agricultural investment maintain the current growth rates), the self-sufficiency rate of maize, soybean, oil, sugar and dairy and other agricultural products will decrease significantly (Table 1-2)[14,15]. The domestic production increase can not meet the growth of domestic demand, and to 2050, 30-60% of the domestic demand for these agricultural products will rely on imports. Although rice supply is more than self-sufficient, and the self-sufficiency rate of wheat can reach 90% or more, the significant increase of the demand for maize may leave its self-sufficiency rate to around 70% by 2050. The low self-sufficiency rate of cotton was mainly due to the expansion of export-oriented garment and textile industries. However, the projection also shows that under the high investment scenario, food security or supply of the agricultural products mentioned above can elevate remarkably. Therefore, the security of domestic food and major agricultural products will be greatly impacted by the agricultural investment and technological changes and it is important to strengthen the study of agricultural science and technology.

Table 1-2 The Self-sufficiency Level of China's Major Agricultural Products, 2004-2050 (%)

	Year			
	2004	2020	2030	2050
Three major cereal	103	93	90	85
Rice	101	102	104	104
Wheat	99	94	92	90
Maize	107	84	79	71
Soybean	49	41	39	38
Edible oil crops	67	62	60	58
Cotton	85	71	64	58
Sugar	91	85	79	75
Vegetable	101	104	105	106
Fruit	101	106	105	104
Pork	101	102	100	98
Beef	100	94	89	84
Mutton	99	94	92	89
Poultry	100	104	105	105
Milk and dairy	96	87	84	79
Aquatic products	102	103	104	104

Source: Huang J K. China's Agriculture toward 2050. A report submitted to IAASTD.Center for Chinese Agricultural Policy, Chinese Academy of Sciences. Beijing. 2008.

Our projections also show that the self-sufficiency level of agricultural products with comparative advantage will maintain a high level in China even in 2050. Although the security of these agricultural products will not face serious threats, the manyfold increase in the demand and production of livestock and aquatic products will pose enormous pressure on agro-ecological issues and pollution caused by quality safety of agricultural products, particularly animal products, aquatic products, vegetables and fruit, food safety and production expansion.

By 2050, food safety and food security issues will be closely watched. A tremendous renovation on agricultural safety control in the production process and quality improvement will be just a matter of time; even domestically, with rising incomes and changing living standards, the demand for food pattern change and quality improvement will continuously upgrade. Agricultural development in China will face greater challenges to ensure the safety and security of agricultural products in the process of economic globalization. Economic globalization and trade liberalization is an inevitable trend. Chinese agriculture will have to face and adapt to the international economic environment with a positive attitude. Under this environment, the security and safety of agricultural products in China will face enormous challenges.

2. Agricultural development in China will face increasingly serious challenges to protect the safety and security of land and water resources

Industrialization and urbanization will be inevitable trends in economic development and social civilization development; however, the limited arable land will be occupied continually under this tendency. As the most active growing economies in the world, China will remain a rapid economic growth in the long term, and urbanization will continue to advance rapidly. Industrialization and urbanization will continue to occupy the limited available land. Meanwhile, to change the basic pattern, "local improvement, but overall deterioration", of eco-environment in China, ecological land will continue to increase in the future. The potential for the development of reserved arable land in China is extremely limited. By 2050, agricultural development in China will increasingly face serious resource constraints.

The projection from Center for Chinese Agricultural Policy states that the measures to significantly expand cultivated area by reclamation to increase agricultural output from the 1950s to early 1980s have become history and will not occur much in the future. Even if the overall cultivated land did not reduce significantly, the total sown area of the three main cereals will continue on a downward trend that started in the late 90s, and this cultivated land area is expected to reduced to less than 660 million mu (15mu=1 hectare) by 2050, down by 13% compared to 2004 of which, except for a steady growth of the area of maize, rice and wheat area will decrease by 25-30% to 2050. In addition, soybean and oil crop sown area will also reduce gradually, but vegetables and fruit area will further expand as a result of the demand-pull reason.

Comparing the projections of sown area for the main crops (Table 1-3) [14,15] with the increase in the demand for agricultural products (Table 1-1) in the future, the increase of agricultural production in the future will mainly come from the increase in crop yield. Therefore, in order to ensure the national agricultural security, agricultural technological breakthroughs are essential.

Table 1-3 The Area Projection for Major Agricultural Crops in China, 2004-2050 (Unit: 10,000 hectares)

	Year			
	2004	2020	2030	2050
Three major cereal	7545	7125	6951	6587
Rice	2838	2425	2263	1964
Wheat	2163	1891	1782	1598
Maize	2545	2809	2906	3025
Soybean	958	889	830	714
Edible oil crops	1443	1363	1366	1363
Vegetable	1756	1833	1894	1973
Fruit	977	1041	1125	1251

Source: Huang J K. China's Agriculture toward 2050. A report submitted to IAASTD.Center for Chinese Agricultural Policy, Chinese Academy of Sciences. Beijing. 2008.

China will face increasingly serious water shortage crises, which will threaten the sustainable development of agriculture. Water crises have been deemed as the biggest factors to threaten global food and agricultural development by the United Nations, the World Bank and other international agencies. As one of the countries whose water shortage is the most serious, the problem in China is particularly acute. With the rapid economic growth, accelerated process of urbanization and industrialization deterioration of water pollution, and the increasing water demand for ecological protection, the threat from water shortages will become more severe for agriculture. The proportion of water for agricultural use has declined from 97% in 1949 to 65% in 2004, and will decline further in 2050 to 40% or even less than 30%. In many regions of China, the rapid industrial development and the rapid increase in water demand from urban population expansion along with income growth has formed fierce competition for limited water resources with agriculture, which threatens food production and food security [16].

3. Agricultural development in China will increasingly be constrained by small-scale production, and the process of agricultural modernization will face enormous challenges

The conflict between small-scale agricultural production and increasing farmers' income, and the conflict between farmers' decentralized operation and modern agriculture and food safety have become increasingly conspicuous. The existing small-scale agricultural production can no longer support the farmers to increase income and output in China. The agricultural production system in China is composed of 240 million small farmers, and average cultivated land per household decreased from around 0.8 hectares in the early 1980s to less than 0.6 hectares in 2007, with a trend of fragmentation. It is almost unlikely to increase farmers' agricultural production initiative and realize sustainable income and output increase by such a small scale of production; consequently, the trends of farmers engaging in agricultural and non-agricultural undertakings simultaneously, farmers no longer mainly living on agricultural undertakings and the aging of rural labor force are becoming increasingly prominent.

A national representative survey conducted by Center for Chinese Agricultural Policy in 6 provinces shows that increasingly more youth in farm villages will not choose agriculture as a major employment opportunity. The survey showed that the proportion of 16 to 25-year-old rural labor force engaged in non-agricultural undertakings has increased from 29% in 1990 to 86% in 2006. The existing small-scale production, part-time farming and agricultural aging have begun to affect the agricultural comparative advantage in China and impede the process of agricultural modernization. Consequently, the initiative of farmers to adopt new technology has dropped, the function of agricultural mechanization has been constrained, and the improvement of agricultural productivity has been impeded. To expand the scale of agricultural production, in addition to the introduction of a range of agricultural and rural development policies (such as land policy, labor market policies, financial policies of credit, agricultural input policy, etc.), constant innovation of agricultural science and technology is required to reduce the cost of agricultural production and promote the development of modern agriculture. The increasing demand for safe and traceable food has presented challenges to the current agricultural production techniques, small-scale agricultural production and farmers decentralized operation.

4. The expansion of domestic and foreign demand for agricultural biomass energy and the increased demand for multifunctional agriculture will further threaten food security and increase the overload pressure on land and water resources

Overall energy is extremely scarcity in China, and the supply and demand deficit is prominent. Especially since China's economy is in the process of continually growing, the contradiction between the domestic energy supply and demand will become increasingly acute. In facing risk of global energy crisis, the major economies in the world (like the United States and EU member states and other developed countries as well as Brazil and other developing countries) have put the development of biomass sources into the main 21st century agenda, and currently, the debate on agricultural biomass energy may not change its long-term trends. China is now urgent for the development of a modern bio-energy industry, and its development will bring about a huge demand for some agricultural products and will present new challenges to the balance of domestic agricultural supply and demand. The huge potential demands for the development of a modern bio-energy industry in China will not only need regular stimulation techniques to promote the overall increase of agricultural products to ease the pressure on products supply, but it also needs to carry out pioneering research and develop energy plants to increase the supply of biomass feedstock in the production.

In addition to providing space for agricultural development, the increased demand for agricultural multifunction will further increase the overloading pressure on land and water resources. Over the past 50 years, China has made great achievements in agricultural development, but also paid a high environmental cost. The land resource base of agricultural survival is deteriorating, and water pollution problems have become increasingly serious. The increased demand for agricultural multifunction such as ecological system improvement, climate change mitigation, and pastoral landscape of agricultural tourism will present enormous pressure on agricultural inputs, technology and the future use of land and water resources.

5. By 2050, agriculture will calmly face the impact of global climate change, and the climate change will bring about many highly uncertain impacts on agricultural development in China

By 2050, global climate change will threaten human survival and ecological systems, and also bring about uncertain factors for agricultural development in China. The China Climate Assessment Report preliminary results show that climate change will be further accelerated in the future in China, and the average temperature is likely to raise 2-3℃ in the next 50-80 years; climate warming is likely to decrease the runoff of rivers to the north, while southern runoff will increase. The average annual evaporation in each basin will increase, and the frequency of droughts and other disasters will increase, which will also increase and exacerbate the instability of water resources and the supply and demand contradiction; climate warming will increase agricultural water demand, and regional differences of water supply will also increase. To meet the changing conditions of production, the demand for agricultural cost and investment will increase substantially; by 2030, China coastal sea levels may rise 0.01–0.16 m, increasing the overflowing risk in coastal regions and posing new threats to marine resources and marine biodiversity. Overall, climate change will bring about significant and uncertain impacts on agricultural production, increasing the instability of agricultural production and enlarging production fluctuations. If no measures are taken now, by the second half of 21st century climate change could lead to significant decreases in the output of major crops such as wheat, rice and maize; during the next 20 to 50 years, if no precautions are taken to formulate responses and make good, sufficient technical reserves, the serious impact from climate change is likely to severely affect food security in China in the long term.

1.3 Outlook for Demand of Major Agricultural Technology in China to 2050

To explore the trend of global agriculture and the foreground of China's agriculture by 2050, China needs to fully utilize the worldwide development opportunities that agriculture faces, and actively respond to the various challenges for global and China's agriculture. China is urgent to make a series of breakthroughs for agricultural development in the following significant technology fields: the utilization of plant germplasm resources and modern breeding; the utilization of animal germplasm resources and modern breeding; the management and safety of resources, agricultural production and food safety; the modernization and informative construction for agriculture.

Realizing the above significant innovations and breakthroughs in crucial technology fields by 2020 China's agriculture must acquire the following diversified technology: the technological supporting qualification for basically improving security of gross national food and fiber supply; the technological supporting qualification for basically ensuring national food safety; the technological supporting qualification for essentially converting the structure of agricultural production; the technological supporting qualification for sustainable utilization and a certain safety improvement of agricultural resources.

By 2030, the technological supporting qualification required in the early stage of new agriculture, which is the transition between traditional agriculture and the combination of traditional and modern multifunctional agriculture, is going to be supplied. It is based on the preconditions: further improvement in sustainable use of agricultural resources and basically guaranteeing security and safety of national food and fiber supply.

By 2050, China should plan to acquire the technological supporting qualifications for the following aspects: resources sustainable utilization; adequately ensuring food security and food safety in nationwide; agriculture entering the new era of existing the traditional function and modern multifunction simultaneously in the future.

Therefore, China is urgent to establish the development goals and specific roadmaps in the following five agricultural technology fields to 2050 (Fig. 1.7). The general development goals of five agricultural technology fields are briefly discussed below:

1) Plant germplasm resources and modern plant breeding science and technology: It is to mainly use the methods of system biology and the products-oriented strategy, exploring the key and functional genes in germplasm and utilizing the superiority of gene resources-rich in China, and making breakthrough in plant photosynthesis research; The researches are to develop genomic-knowledge based on key biotechnology, and constructing the new innovation system of functional plant products development to provide the scientific supporting to agricultural sustainable development.

2) Animal germplasm resource and modern breeding science and technology: It is mainly to use integration of multidisciplinary research methods in the fields of life science and biotechnology such as systems biology, bioinformatics, genomics and proteomics, genetic engineering etc. It adopts the major product-oriented research & development strategy to develop healthy and sustainable development of animal aquaculture, including marine fisheries, on the purpose of cultivating livestock, poultry and aquatic products, which are safety, quick growth, high protein-content, high meat yield, high feed transformation or resistance to diseases.

3) Resource saving agricultural science and technology: It is to develop high-efficient utilization mechanisms and methods on water, nutrients, and artificial assistance resources. Main investigations focus on arable land nourishment and substitution technology, engineering-biology-chemistry water saving technology, watershed water resource management technology, water-fertility-energy integrated management, precision mechanized implement technology, and intelligent fertilizer. It is to establish three agriculture production technology systems, e.g., land-saving agriculture, water-saving agriculture, fertilizer- and energy-saving agriculture. It is to realize agriculture production intensification, mechanization, large-scale operation and industrialization. It is to realize the enhancement of sustainable agriculture production and the utilization of resources.

4) Agricultural production and food safety science and technology: It is to achieve the breakthroughs of basic theories and key technologies on agricultural production safety, prevention and control of major pests and diseases; storage, retaining freshness, food processing and food nutrition, etc., the establishment of the prevention and control warning system of agricultural pests and diseases, implementation of accurate monitoring and prevention and control of "active security strategy". It is to establish a standardized system of safe production, distribution & preserve and processing for agricultural products on purpose of providing products with food safety and high-quality guarantee. It is to establish precise food design and the various nutrition food system to meet the nutritional materials which human health requires for. And it is also to create "intelligent personalized nutritional food" system to meet the physiological properties and health requirements by various groups of people and provide personalized functional foods.

5) Agricultural modernization and intelligent agricultural science and technology: It is to realize the following goals through the breakthrough of the key technologies and equipped by high-tech. These goals are: agricultural information service network, digital management of agricultural resources, precision management of the agricultural production process, and intelligent of agricultural equipment.

In this report, chapter 2 is going to summarize the general-goal and subgoal of China's agricultural science and technology development, and the realizing general roadmap for them.Chapter 3 and 7 will introduce the general-goal, sub-goal and the particular roadmaps for realizing the goals in each of five science and technology fields which including: plant germplasm resources and modern breeding, animal germplasm resources and modern breeding, saving resources, agricultural production and food safety, agriculture modernization and intellectualized agriculture. Chapter 8 will bring forward the institution and significant policy implications for implementing the roadmap of China's agricultural science and technology development.

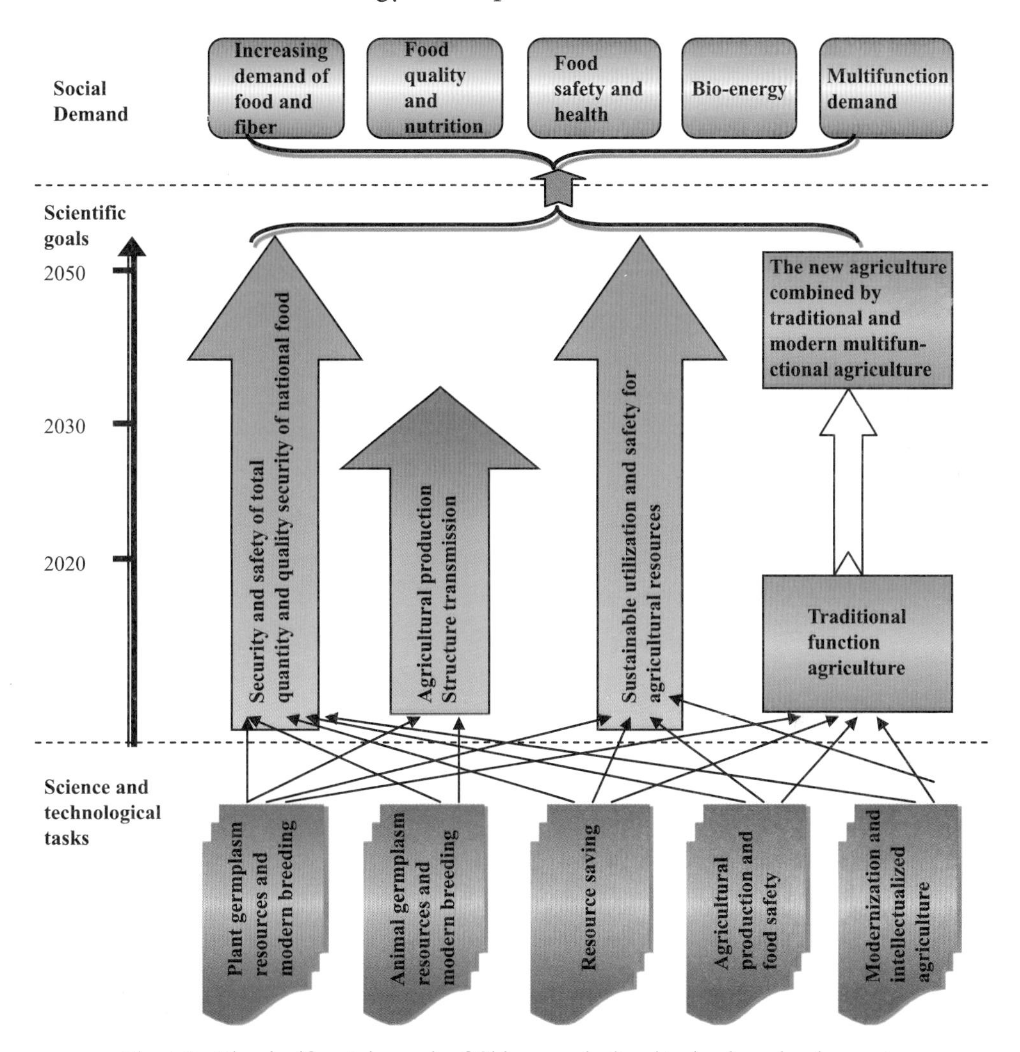

Fig. 1.7 The significant demands of Chinese agricultural technology development

Main References

[1] UN (the United Nations).UN Population Division, Department of Economic and Social Affairs, UN Secretariat. New York .World population prospectus: The 2004 revision. 2005.

[2] IFPRI (International Food Policy Research Institute). Washington DC. IMPACT simulations: Global Agriculture toward 2050. 2008.

[3] FAO (Food and Agricultural Organization). Rome.Soaring Food Prices: Facts, Perspective, Impacts and Actions Required. High-Level conference on World Food Security, 2008.

[4] OECD. Paris. Economic Assessment of Biofuel Support Policies. Directorate for Trade and Agriculture, 2008.

[5] RFA (Renewable Fuels Associations). Ethanol Industry Outlook. [2008. Various Issues]. http://bioconversion.blogspot.com

[6] Yang J, Qiu H G, Huang J K, et al. Fighting Global Food Price Rises in the Developing World: the Response of China and Its Effect on Domestic and World Markets. Agricultural Economics, 2008, 39: 453-464.

[7] US Renewable Fuels Association [2008]. http://www.ethanolrfa.org/industry/statistics/

[8] Earth Policy Institute. World Ethanol Production and World Biodiesel Production [2008]. http://www.earthpolicy.org/Updates/2006/Update55.

[9] BIODIESEL 2020.Global Market Survey: Feedstock Trends and Forecasts [2008-02]. http://www.emerging-markets.com/biodiesel.

[10] Huang J K, Yang J, Qiu H G, et al. Development and Impacts of Global and GMS Regional Biofuels in Agriculture and the Rest of Economy with Specific Focus on the GMS. A Report submitted to Asian Development Bank. Manila. 2008.

[11] IAASTD (International Assessment of Agricultural Knowledge, Science and Technology for Development). Agriculture at A Crossroads: Global Report. Washington: Island Press. 2008.

[12] Huang J K, Rozelle S, Pray C. Enhancing the Crops to Feed the Poor. Nature, 2002,418 (8): 678-684.

[13] IPCC (Intergovernmental Panel on Climate Change). Climate Change 2001: Synthesis Report. Cambridge: Cambridge University Press. 2001.

[14] Huang J K. China's Agriculture toward 2050. A report submitted to IAASTD.Center for Chinese Agricultural Policy, Chinese Academy of Sciences. Beijing. 2008.

[15] National Bureau of Statistics of China. China Statistical Yearbook. Beijing: China Statistics Press. 2004-2008.

[16] Wang J X, Huang J K, Rozelle S, et al. Understanding the Water Crisis in Northern China: What the Government and Farmers are Doing. Water Resources Development, 2009, 25(1) : 141-158.

2 Roadmap and Goals of Agricultural Science and Technology in China

2.1 The General Goal

In the next 50 years, the overall goal for agricultural science and technology in China is to ensure the sustainable development of resources, environment and socio-economy, and on this basis to provide feasible and innovative agricultural science and technology support and security system for the development of agricultural products to meet the continual increase and change in demand (traditional or non-traditional) for agricultural products from human society.

In the field of plant germplasm resources and modern plant breeding, using the methods of system biology mainly and the products-oriented strategy, exploring the key and functional genes in germplasm and utilizing the superiority of gene resources-rich in China, making breakthroughs in plant photosynthesis research and developing genomic-knowledge based key biotechnology, constructing the new innovation system of functional plant products development, improving the yield potential, quality, and exploring multifunctional and smart cultivars, providing the scientific supporting to agricultural sustainable development.

In the field of animal germplasm resource and modern breeding science and technology, we mainly use integration of multidisciplinary research methods in the fields of life science and biotechnology such as systems biology, bioinformatics, genomics and proteomics, genetic engineering, etc., with major product-oriented research & develop strategy, develop healthy and sustainable development of animal aquaculture, including marine fisheries, cultivate livestock and poultry seafood with safety, growth, high protein content, high meatyield, high feed transformation or resistance to diseases, Fully explore and use our abundant resource of aquatic animals.

In the field of resource saving agricultural science and technology, it is to perfect the monitoring and disaster-forecasting platform of national arable land and water resources, to perfect management the technology research platform of water and nutrients, and to perfect the new fertilizer research and development platform. It is to establish three agricultural production technology systems including land-saving agriculture, water-saving agriculture and fertilizer- and energy- saving agriculture.

It is to enhance a new fertilizer industry and the modern agricultural equipment industry. It is to realize intensified utilization and management of regional water and soil. It is to implement agricultural mechanization by way of water-saving irrigation and highly-efficient fertilization. It is to standardize precision water and fertilizer management and energy-saving minimal and zero tillage. It is to secure that arable land area with medium and low yield should be reduced by 50% to 60 %. The comprehensive utilization rate of soil, fertilizer and water in agriculture ecosystem should be improved by 30%. The nutrient and energy input should be reduced by 25% to 30%. It is to apply intelligent fertilizer widely, to realize the dynamic balance between arable land and water resources which food security production depends on. It is to set up modern agricultural production system with a high and stable yield, high efficiency and high quality in order to realize sustainable agriculture.

In the field of agricultural production and food safety science and technology, it is going to enhance the basic theoretical researches on agricultural production safety, prevention and control of major pests and diseases; maintain the nutrition of agricultural products, clean control, storage and processing and so on, the key technological breakthroughs and integrated technologies integrative innovation, the establishment of the prevention and control warning system of agricultural pests and diseases, intelligent expert management system, and form the food safety digital tracing system from farm to fork, implementation of accurate monitoring and prevention and control of "active safety strategy". It is planning to establish a standardized system of safe production of agricultural products and green environment, and to ensure the quality and safety of agricultural products in the cultivation, breeding, storage and processing. It is to achieve precise design and quality regulation of food, and proceed the various nutrition food R&D to improve the diet combination, and to create "intelligent personalized nutritional food" to meet the personalized nutritional and healthy needs.

In the field of agricultural modernization and intelligent agricultural science and technology, it is to establish a modernization platform for promote innovations on Chinese agricultural researches and realize the following goals through the breakthrough of the key technologies and equipped by high-tech. These goals are: agricultural information service network, digital management of agricultural resources, precision management of the agricultural production process, intelligent of agricultural equipment and agricultural machine, network platform for virtual research in agriculture, rapidly increasing agricultural productivity, resources efficiency and agricultural continuous innovation ability.

2.2 Target Phases

The development of agricultural science and technology is including three phases that are short-term (2020), mid-term (2030) and long-term. The target for each phase of agricultural science and technology development is showing in the followings (Table 2-1).

Table 2-1 The Summary Table for Targets Phases of Agricultural Science and Technology Development

Scientific assignments	2020	2030	2050
Plant germplasm and modern plant breeding	① Establishing the gene bank and database for ecological populations and germplasms and its information sharing-system ② Developing multigene transgenic and multitraits improved cultivars ③ Realizing the high-efficiency transformation and congregation of major-effect gene and its interacting network	① Drawing the distributing map of germplasm resources and dynamic map of ecological populations ② Obtaining primarily the new crops by design and assembling, breeding and releasing new super-energy plant cultivars	① Developing new smart plant cultivars by combing the technology of molecular design and genome-based bioinformatics ② Developing and releasing a large number of high-energy content plant cultivars for industry of bio-energy
Animal germplasm resource and modern breeding	① Establishment of the animal germplasm resource sharing platform of important aquatic animal ② Development of molecular markers and explore germplasm resource of special value ③ To identify the major genes of economic traits and reveal the gene regulatory networks and the major gene	① Establishment of molecular design breeding of important livestock, poultry and aquatic animals ② Establishment of sex control breeding, disease resistance breeding, intelligent molecular design breeding, multifunction molecular design breeding techniques ③ To unearth a crop of important functional genes of livestock, poultry and aquatic animals	① Analysis of the major pathogens through pathogen biology, pathogenesis, and epidemic law, develop effective ecology strategies to control aquatic animal disease, develop suitable vaccines and effective drugs ② Development of effective drugs with high value of output and no harm to human health
Resource saving agriculture	① Establish monitoring and disaster forecast platform for arable land and water resources, platform for water and fertilizer utilization research, and platform for new-type fertilizer research and development ② Establish arable land coordinating and coupling utilization and soil quality directive breeding technologies ③ Research on water-saving irrigation engineering technology and its integrating soil tillage, water saving and straw cover technology ④ Research on fertilizer saving and zero tillage technologies as well as new slow/controlled released fertilizer. ⑤ Medium and low yield land area down by 30-40 %, utilization rate of water and fertilizer up by 10 %, nutrient resources input down by 15-20 %, apply compound fertilizer and slow released fertilizer widely	① Establish arable land substitution technology and constrained soil restoration technology ② Research on various water saving technology and management of river basin water resources ③ Research on mechanized technology of water-fertilizer-energy coupling management and multi-functional biological compound fertilizer ④ Medium and low yield land area down by 40-50%, utilization rate of water and fertilizer up by 20%, nutrient resources input down by 20-25 %, apply slow/controlled released fertilizer and multi-functional fertilizer widely	① Establish large-scale operation and management system of regional arable land and high-yield soil nourishing technology ② Establish all-round and high-efficient utilization system of river basin water resources and engineering-biology-chemistry water-saving integrated technology ③ Establish comprehensive management system of regional agricultural nutrient resources, and develop intelligent fertilizer ④ Medium and low yield land area down by 50-60%, utilization rate of water and fertilizer up by 30%, nutrient resources input down by 25-30%, apply intelligent fertilizer widely

Continued

Scientific assignments	2020	2030	2050
Agricultural production and food safety	① Strengthen the basic research on the related scientific issues according to the key factors that affect the quality and safety of agricultural products ② With the application of breakthroughs in key technologies and the integrated application of comprehensive technologies to eliminate the risk factors that affect the agricultural products safety, establish the standardized technical system of agricultural safety, storage and processing ③ Construct the ecological environment of green food production and provide green, safe and high quality food	① Reveal the regulatory mechanism of factors that affect agricultural products safety, promote the standardization and safe production mode of agricultural products, and realize the precise control of storage, logistics and processing quality of agricultural products ② Clarify the principles of restoration of contaminated environment, establish the management system of intelligent environmental monitoring and remediation of contaminated soil, create a sustainable ecological environment ③ Study the theory of agricultural nutrients, use the accurate and rapid intelligent design platform to develop processed foods with various nutrition and provide a variety of nutrients that human health needed	① Clarify the metabolic and synthetic way of various nutritional components, develop a harmonious agricultural environment which can satisfy multi-needs, lay the theoretical foundation of quality regulation of processed foods ② Set up the design standards of personalized nutritional foods, create "intelligent personalized nutritional food" based on the physiological and health characteristic of different groups, meet the individual nutritional needs, improve the physique and health level of all the people and provide individualized functional foods to effectively prevent and reduce the diseases
Agricultural modernization and intelligentization agriculture	① Complete the development of multi-functional agricultural information network platform and professional search engine ② Complete agricultural information service network convergence ③ Complete the key technology development for precision, intelligent, digital, virtual agriculture	① Complete the development of multi-functional agricultural information network platform and professional search engine ② Realize the convergence of precision, intelligent device and system during the plant and cultivation, and implement the large-scale demonstration and promotion ③ Complete the construction of resource digital management system in national, provincial, municipal and county levels and implement the digital management promotion ④ Complete the construction of the network platform for agricultural virtual research	① Realize the agricultural resources digital management basically ② Complete the development of a series of intelligent equipment and proceed the practical applications ③ Conduct the precise management to animals and plants during the production process and realize comprehensively agricultural informationization and precise management ④ Realize the agricultural research innovation based on the agricultural virtual research platform

The goals of agricultural science and technology development to 2020 mainly are: establishing the gene bank and database for ecological populations and germplasms and its information sharing-system, developing multigene transgenic and multitraits improved cultivars, perfecting the technology of genetic transformation and elite germplasm innovation in main crops, combing with molecular marker assisted selection and safe transformation technology, realizing the high-efficiency transformation and congregation of major-effect gene and its interacting network; establishing important livestock, poultry and aquatic animals germplasm resource sharing platform, develop molecular marker technology and explore special value of germplasm resource and enhance genetic improvement strength and potential innovation of the fine varieties of livestock, poultry and aquatic animals; it is to develop primarily three agriculture production technology systems, e.g., land-saving agriculture, water-saving agriculture, fertilizer and energy-saving agriculture in order to establish the production base for sustainable agriculture development, the precise management to animals and plants will be conducted during the production process; realize comprehensively agricultural informationization and precise management; establishing the standardized system of agricultural production safety, storage and processing, and it is not only structuring the green environment for green food production, but also providing green safety agricultural products with high-quality; completing the development of multi-functional agricultural information network platform and professional search engine, establishing a large regional scale professional database and agricultural information service network, proceeding the researches on key technologies of digital resource management and precision management of the agricultural production process, intelligent of agricultural equipment and prototype of the network platform for agricultural virtual.

The five main goals of science and technology field to 2020 are presenting in the next paragraphs.

The goal should be achieved in the field of plant germplasm resources and modern plant breeding is, establishing the gene bank and database for ecological populations and germplasms and its information sharing-system, developing multigene transgenic and multitraits improved cultivars by combing the traditional and modern technology of plant breeding, perfecting the technology of genetic transformation and elite germplasm innovation in main crops, combing with molecular marker assisted selection and safe transformation technology, realizing the high-efficiency transformation and congregation of major-effect gene and its interacting network, breeding new energy plant cultivars with wide adaptation ability.

The goals to be achieved in the fields of animal germplasm resource and modern breeding science and technology are: to explore particular germplasm recourse of important aquatic animals in our country, to establish animal germplasm resource sharing platform and protect animal germplasm resource of important aquatic animals, to develop molecular marker technology and explore special value of germplasm resource, to enhance genetic improvement strengths and potential innovation of the fine varieties of livestock, poultry and aquatic animals.

The goals will be reached for resource saving agriculture science and technology in 2020 as following: It is to set up the monitor and disaster forecast platform of arable land and water resources, comprehensive management and research platform of water, fertilizer and tillage as well as new high-efficient fertilizer research and development platform at regional scale. It is to implement intensified utilization of regional arable land resources and land saving technology, and at the same time combine oriented nourishing technological system for arable land fertility with an aim to stabilizing arable land area and its quality. It is to implement comprehensive management policies and measure of river basin water resources in case of climatic changes, and realize stable supply and reasonable utilization of river basin water resources. It is to implement water-saving irrigation, minimal and zero tillage technology, farm straw transforming technology as well as high-efficient fertilization technological system. It is also to work on a series of highly efficient compound fertilizer and new controlled release fertilizer, as well as agricultural tillage and fertilization machines. It is to develop new species technology, to set up regulations of agriculture cultivation, to reduce the input of water, fertilizer and energy overall. Finally, they can save the cost of agricultural production. It is to guarantee that medium and low yield land area should be reduced by 30% to 40%. The comprehensive utilization rate of soil, fertilizer and water of farm ecosystem should be improved by 10%. The nutrient and energy input should be down by 15% to 20 %. It is to apply compound fertilizer in grain crops widely, and to begin to apply slow released fertilizer for cash crops such as vegetables and fruits. It is also to stabilize the quantity and quality of the arable land at national scale. It is also to set up three agricultural production technological systems, e.g., land saving agriculture, water saving agriculture, fertilizer and energy saving agriculture. They can lay the foundation for sustainable development of agriculture.

The goals should be achieved in the field of agricultural production and food safety science and technology are: enhancing the basic research on relevant scientific issues against the main factors which impact the agricultural products safety, eliminating the hazard factors of impacting agricultural products safety through the breakthroughs on key technology and comprehensive applications on integrated technology; establishing the standardized system of agricultural production safety, storage and processing, and it is not only structuring the green environment for green food production, but also providing green safety agricultural products with high-quality.

The goals will be reached in the field of agricultural modernization and intelligent agricultural science and technology are: completing the development of multi-functional agricultural information network platform and professional search engine, establish a large regional scale professional databases, especially resource archive database, model base and refreshable system, complete agricultural information service network, provide services of regional resources optimization, production configuration, technical consulting, market demand analysis, product traceability, etc., study key technologies of digital resource management, precise management in manufacturing process, intelligent agricultural equipment, obtain breakthrough in remote sensing of disaster monitoring, drought monitoring, crop growing monitoring, soil moisture information collection, precise seeding, precise irrigation, etc., study the prototype of the network platform for agricultural virtual research, identify the feasibility of the platform by carrying out the demonstration application on typical subject.

The goals of agricultural science and technology development to 2030 mainly are: perfecting the technology of molecular design, hasting the technology of plant breeding by molecular design, developing and releasing new cultivars for crop production. Obtaining primarily the new crops by design and assembling, breeding and releasing new super-energy plant cultivars; establishing the molecular design of livestock, poultry and aquatic animals breeding technology with Chinese characteristics and breed new varieties of pig, cattle, sheep, chicken ,fish, shrimp, shellfish and other key breeding animals with quick growth, high protein content, high meatyield, high feed transformation or resistance to diseases; perfecting three agriculture production technology systems, e.g., land-saving agriculture, water-saving agriculture, fertilizer and energy-saving agriculture in order to realize leapfrog development for agriculture production intensification, mechanization, large-scale operation and industrialization; realizing the precise control of storage, logistics and processing quality of agricultural products, establishing the management system of intelligent environmental monitoring and remediation of contaminated soil, creating a sustainable ecological environment, and developing processed foods with various nutrition and provide a variety of nutrients that human health needed, realizing the agriculture information service network around China, quantitative management of soil, water and weather resources, dynamic monitoring of farmland water, pest and weed, intelligent precise management of farmland water, pesticide, cultivation and breeding, and establishing the network platform for agricultural virtual research.

The five main goals of science and technology field to 2030 are presenting in the next paragraphs.

The goals to be achieved in the field of plant germplasm resources and modern plant breeding are: drawing the distributing map of germplasm resources and dynamic map of ecological populations, perfecting the technology of molecular design, hasting the technology of plant breeding by molecular design, developing and releasing new cultivars for crop production. Obtaining primarily the new crops by design and assembling, breeding and releasing new super-energy plant cultivars.

The goals to be achieved in the fields of animal germplasm resource and modern breeding are: to explore functional genes and large-scale cloning with our own intellectual property rights, to establish the molecular design of livestock, poultry and aquatic animals breeding technology with Chinese characteristics, to breed new varieties of pig, cattle, sheep, chicken, fish, shrimp, shellfish and other key breeding animals with quick growth, high protein content, high meatyield, high feed transformation or resistance to diseases; together with traditional breeding methods, to establish sex control breeding, disease resistance breeding, molecular design of intelligent animal breeding, molecular design of multi-trait breeding techniques for new varieties.

The goals will be reached for resource saving agriculture science and technology development in 2030 as following: It is to realize the arable land substitution technology, high yield arable land conservation technology, medium and low yield land restoration technology, and to implement all-round management of constrained soil in realization of the stability and enhancement in terms of both quantity and quality of arable land. It is to set up safeguard policies and technological system for river basin water resources, and secure the dynamic balance of water and soil resources which the food security production depends on. It is to work all-round on new multi-functional fertilizers for different regional climate and cultivation conditions, to set up new multi-functional fertilizer industry, to develop a series of slow/controlled released fertilizer and multi-functional fertilizer, and to do research on new technological system for biomass cultivation and transformation. It is to implement precision management of water and fertilizer and energy-saving cultivation integrating technological system. In coordination with the development of light energy utilization technology, it is to set up new and high-efficient planting system. It is also to set up water-, fertilizer- and energy-saving agricultural production system and enhance the utilization rate of agricultural resources. It is to guarantee that medium and low yield land area should be reduced by 40% to 50%, the comprehensive utilization rate of soil, fertilizer and water of farm ecosystem should be improved by 20%. The nutrient and energy input should be down by 20% to 25%. It is to apply widely compound fertilizer and low-cost and slow released fertilizer for grain crops, and to begin to apply controlled released fertilizer and multifunctional biological fertilizer for cash crops including vegetables and fruits. It is to enhance the quality of arable land resources steadily and improve three major agricultural production technological systems of land saving agriculture, water saving agriculture, and fertilizer and energy saving agriculture. They can realize leapfroging development of agricultural production in terms of intensification, mechanization, standardization, and industrialization.

The goals to be achieved in the fields of agricultural production and food safety science and technology are: revealing the regulatory mechanism of factors that affect agricultural products safety, promoting the standardization and safe production mode of agricultural products, and realizing the precise control of storage, logistics and processing quality of agricultural products; clarifying the principles of restoration of contaminated environment, establishing the management system of intelligent environmental monitoring and remediation of contaminated soil, creating a sustainable ecological environment; studying the theory of agricultural nutrients, using the accurate and rapid intelligent design platform to develop processed foods with various nutrition and providing a variety of nutrients that human health needed.

The goals to be achieved in the fields of agricultural modernization and intelligentization science and technology are: finishing the construction of the large-scale professional database in whole country, realizing the agriculture information service network around China; making breakthroughs in small scale nutrient information acquisition technology, remote sensing of crop quality monitoring, the growth simulation of animals and plants, variable fertilize equipment; realizing quantitative management of soil, water and weather resources; realizing dynamic monitoring of farmland water, pest and weed, realize intelligent precise management of farmland water, pesticide, cultivation and breeding; establishing the network platform for agricultural virtual research to provide high-power service and cooperation environment for evolving a series of cross-regions, cross-organizations and cross-subjects agricultural scientific research activities in many fields, such as animal and plant breeding, high-yield cultivation, forecast and monitoring of pathogen and pest, the monitoring and assess of soil quality and weather forecast.

The goals of agricultural science and technology development to 2050 mainly are: realizing genome-wide gene optimization and assembling, developing smart new plant cultivars with high yield potential, good end-using quality and multiple functions, developing and releasing a large number of high-energy content plant cultivars for industry of bio-energy; developing effective ecology strategies to control livestock, poultry and aquatic animals disease, and develop suitable vaccines and effective drugs; to discover a crop of effective drugs with high value of output and no harm to human health, and then achieve the ecological management of aquaculture; establishing overall three agriculture production technology systems, e.g., land-saving agriculture, water-saving agriculture, fertilizer and energy-saving agriculture in order to stabilize sustainable agriculture development; developing a harmonious agricultural environment which can satisfy multi-needs, setting up the design standards of personalized nutritional foods, creating "intelligent personalized nutritional food" to meet the individual nutritional needs; completing the construction of varies scale information collection system and realize the agricultural resources digital management basically, conducting the precise management to animals and plants during the production process, realizing comprehensively agricultural informationization and precise management, and realizing the agricultural research innovation through the network platform of virtual agriculture.

The five main goals of science and technology field to 2050 are presenting in the next paragraphs.

The goals should be achieved in the field of plant germplasm resources and modern plant breeding are: developing smart new plant cultivars with high yield potential, good end-using quality, multiple functions and quickly response to environment stress by genome-wide gene optimizing and assembling, developing and releasing a large number of high-energy content plant cultivars for industry of bio-energy.

The goals are about to be achieved in the fields of animal germplasm resource and modern breeding science and technology are: to analysis the major pathogens through pathogen biology, pathogenesis, and epidemic law, to develop effective ecology strategies to control aquatic animal disease, to develop suitable vaccines and effective drugs; to discover a crop of important functional genes of livestock, poultry and aquatic animals; to develop effective drugs with high value of output and no harm to human health; to achieve the ecological management of aquaculture; to promote health fisheries aquaculture, protect water resource and make marine and freshwater aquaculture with scientific layout, rational structure, abundant variety, complete specifications, high quality and energy saving.

The goals will be reached for resource saving agriculture science and technology in 2050 as following: Based on the improvement of information-based agricultural resource management system, it is to set up resource security and regulation system of water and soil, with an aim to completely realizing the dynamic balance management of arable land and water resources. It is to implement arable land nourishing technology, to control soil degradation processes in different agricultural zones, and to improve the arable land quality steadily. It is to coordinate the development of new species of crops and material technology, and to develop manufacturing technologies of new intelligent fertilizer. It is to apply high-efficient utilization technological system of farm biomass resources. Based on the development of digitized and intelligent agricultural production technologies, it is to implement all-round precision management of water and soil suitable for crop growth as well as planting and tilling management system, realizing precisely mechanized agricultural production. It is to guarantee that medium and low yield land area should be reduced by 50% to 60 %, The comprehensive utilization rate of soil, fertilizer and water of farm ecosystem should be improved by 30%. The nutrient and energy input should be down by 25% to 30 %. It is to apply widely slow released fertilizer and multi-functional biological fertilizer for grain crops, and to begin to apply intelligent fertilizer for cash crops including vegetables and fruits. It is to improve the quality of arable land resources and to establish all-round three major agricultural production technological systems of land saving agriculture, water saving agriculture, fertilizer and energy saving agriculture, and to stabilize the sustainable agriculture development.

The goals are going to be achieved in the fields of agricultural production and food safety science and technology are: clarifying the metabolic and synthetic way of various nutritional components, developing a harmonious agricultural environment which can satisfy multi-needs, laying the theoretical foundation of quality regulation of processed foods, setting up the design standards of personalized nutritional foods, creating "intelligent personalized nutritional food" based on the physiological and health characteristic of different groups, meeting the individual nutritional needs, improving the physique and health level of all the people and providing individualized functional foods to effectively prevent and reduce the diseases.

The goals will be achieved in the fields of agricultural modernization and intelligentization science and technology are: finishing the construction of varies scale information collection system and realizing the agricultural resources digital management basically; completing the development of a series of intelligent equipment and the precise management to animals and plants will be conducted during the production process; realizing comprehensively agricultural informationization and precise management; realizing the wide application of the network platform for agricultural virtual research, system optimization, and implementing the agricultural innovation research.

2.3 General Roadmap

In order to realize the previous general and phase goals, this report sets up the specific roadmap for 5 fields respectively and gathers them together to obtain the general roadmap for agricultural science and technology (Fig. 2.1). This roadmap is including the cores of five key fields and representing as following.

In the field of plant germplasm resources and modern plant breeding, using the combined methods and technology of system biology, bioinformatics, gene engineering, life science, information science and material science mainly and the products-oriented strategy, exploring the key and functional genes in germplasm and utilizing the superiority of gene resources-rich in China, constructing the genebank for strategy-plant in different aspects, optimizing design plant new genotypes based on demands, screening the target genotypes and assembling the ideal genotypes, quantitatively activating or silencing the target genes, developing smart new cultivars of food crops, forage crops and energy crops with high yield potential, good end-using quality, multiple functions and quickly response to environment stress.

In the fields of animal germplasm resource and modern breeding, we fully explore and use our abundant resource of livestock, poultry and aquatic animals, develop germplasm resource evaluation, exploration, conservation and use of technology and enhance the intensity of livestock, poultry and aquatic animal genetic improvement innovative potential. Together with traditional breeding methods, we clone and identify functional genes for molecular breeding design, breed new varieties of pig, cattle, sheep, chicken, fish, shrimp and shellfish with quick growth, high protein content, high meatyield, high feed transformation or resistance to diseases and develop pathogen-specific vaccine with high efficiency, non-disease intensive breeding, provide health and safe production for human.

For resource saving agriculture science and technology, firstly, it is to set up monitor system for arable land, to establish the platform for water, fertilizer and energy utilization, and to build the platform for new fertilizer research and development. Secondly, on the basis of the theories of evolution and adjustment of regional arable land and water resources, cycling and controlling of water, fertilizer and energy of farm ecosystem, degraded soil prevention and restoration, as well as oriented nourishing fertility, it is to realize intensified utilization of water and soil resources, to achieve breakthrough in land saving technology, water-saving irrigation technology, minimal and zero tillage technology, efficient utilization technology of farm biomass resources, as well as efficient fertilizing technological system, to do researches on agriculture supporting facilities, and to standardize production regulations. Thirdly, it is to develop new types of slow and controlled released fertilizer and intelligent fertilizer, and to set up new fertilizer industry. Finally, in light of the development of information-based agriculture and crop species, it is to establish all-round three principal production technological systems of time and land saving agriculture, water saving agriculture, and energy saving agriculture, and to realize intensified production of agriculture and improvement of stable production capacity of arable land for stable development of sustainable agriculture.

In the field of agricultural production and food safety science and technology, it is to fully absorb and use the computer network technology, 3S technology, computer visualization, geographic information system, high-quality satellite images, BP neural network, intelligent expert systems, high-precision technology management, establish the early warning and monitoring system, and intelligent expert management system of animal pests and diseases, and set up the prevention-based comprehensive prevention and control of food safety technology and management system; establish the food safety digital tracking and warning systems from farm to fork, carry out the "active food safety strategy" with accurate monitoring, pre-emptive "disease" and precise prevention and control. Based on the comprehensive analysis of all elements of animal and plant safety products, with the use of the precision and rapid high-throughput based detection and intelligent design platform, develop "intelligent personalized nutritional food", to satisfy all the people's permanent need for food safety and nutrition.

In the field of agricultural modernization and intelligentization technology development, the technology which the agriculture information service network mainly depends on is fairly mature and has the relevant basic conditions, and it should give priority to make progress before 2020. Digital management of agricultural resource, precise management of agricultural production process, intelligent agricultural equipment depend on many high-tech developments, which are still at the research period of exploitation, and need to plan with long-term development. The basic design idea is breaking through the key technologies difficulty first, solving the bottleneck problem, then studying and developing the key component, carrying on the important systemic integration finally.

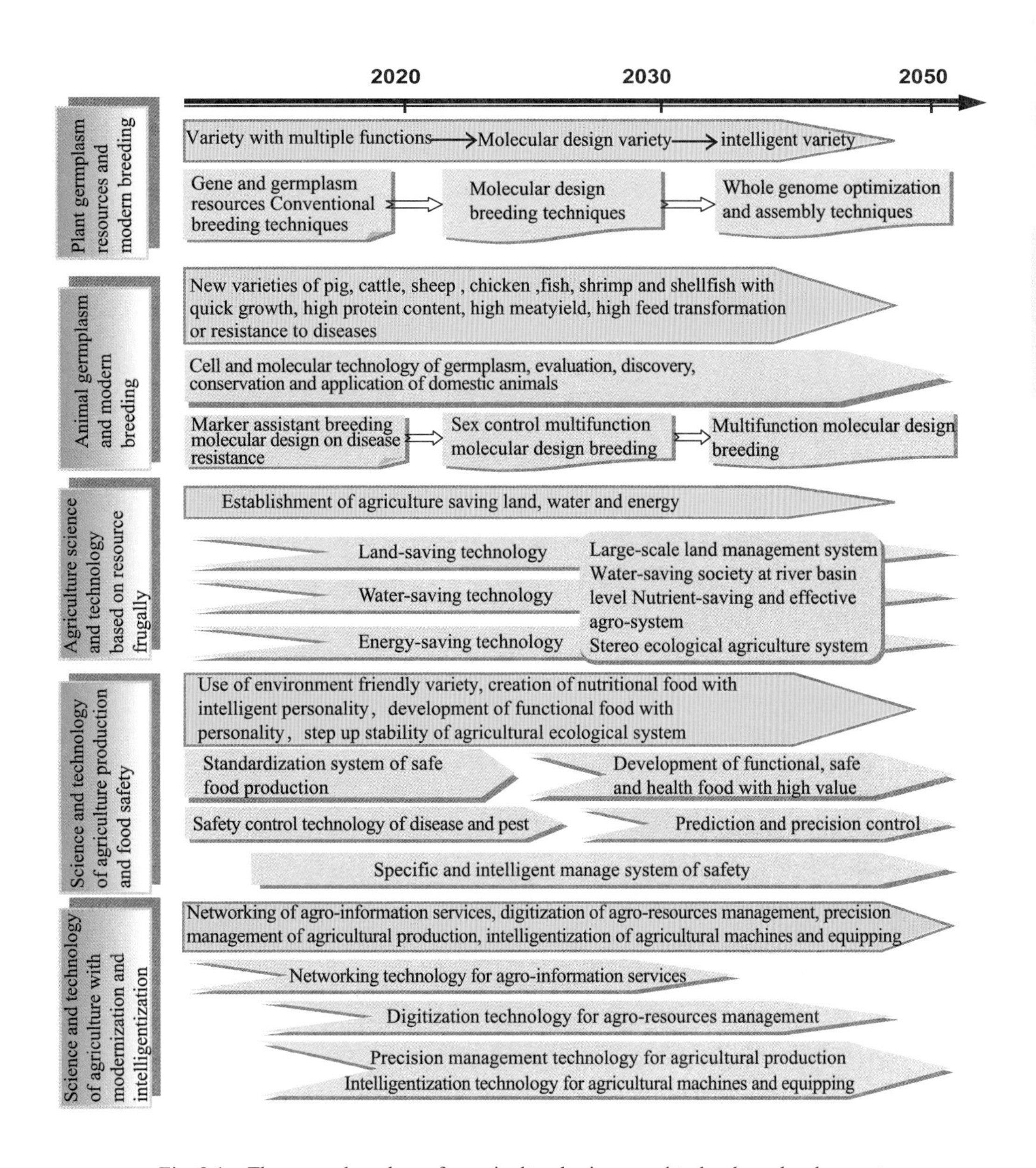

Fig. 2.1 The general roadmap for agricultural science and technology development

3 Roadmap of Plant Germplasm Resources and Modern Plant Breeding Science and Technology Development

Up to 2050, the demand of human on agricultural products increasing not only on total production but also on construction, purpose and quality. In order to guarantee the total food supply and the safety of quality, and provide the support for new development of both traditional agriculture and modern multifunctional agriculture, we need a series of breakthroughs in field of plant germplasm resources and modern plant breeding.

3.1 Requirements and Trends for Future Development

3.1.1 Requirements and Trends for Future Development

Chinese agriculture in the future will face two immediate challenges. Its total production must rise continuously to meet the increasing demands of food and industrial materials. The infrastructure of Chinese agricultural system must undergo rapid transformation to adapt to changes in the patterns of food consumption. In addition, important breakthroughs in crop breeding and production are needed to lessen the constraints imposed by decreasing arable land and water resources. The realization of the huge potential in bioenergy development and consumption needs not only improvement in conventional production technique but also innovative strategy. Finally, significant progress in crop genetic improvement and cultivation measures must be made in order to deal with the uncertainties caused by global climate change. Together, the problems listed above raise new demands in the research on crop resources and breeding science, particularly in the following aspects.

1. Revolutionary crop varieties are needed to ensure not only the adequate supply of food but also the increased development of fiber, oil, sugar, vegetable and ornamental crops

In China, the strenuous relationship between food supply and demand will persist for the foreseeable future. In the mean time, as the living standards rise, consumers pay more and more attention to not only food quality but also the quantity and quality of additional agricultural products. Apart from the progress in current yield promoting techniques, there will be diversified needs in the improvement in crop quality and varieties. On the one hand, the quality of existing food products must be improved substantially. The quality traits related to nutrition, appearance and flavor will be important targets for such improvement. The combination of conventional breeding and genomics based biotechnology will be the most effective approach for achieving the goals in yield, quality trait improvement. On the other hand, the crop varieties producing functional foods are in severe shortage. In addition to basic nutrients, functional crop varieties provide special health promoting components, which may bring certain benefits to the regulation and balance of the metabolic pathways in human bodies. The consumption of functional foods will become an important means for increasing human immunity, preventing diseases and delaying aging in the future. Consequently, the manufacture of functional foods is a new trend in the research and development of food industry. The applications of oral vaccines derived from transgenic plants to prevent animal diseases are attracting more and more investment and in worldwide biotechnology research.

2. The development of bioenergy industry requires strong inputs from multiple technologies involving plant production, resources and breeding

In the future, the dependence of Chinese national economy on plant production will increase continuously, especially with the advent of bioenergy industry. The abundant supply of raw plant materials may become the bottleneck of bioenergy industry, if the plant species that are amenable to large scale cultivation, and can attain very high level of biomass are not available. As biomass accumulation relies on photosynthesis with 90% to 95% of plant dry matter coming directly from photosynthates, significant improvements in the light energy use efficiency of plants are of great importance to biomass production. Theoretically, the light use efficiency of crop plants can reach as high as 5%. But at present, this value stands only around 1% to 1.5% in most rice and wheat varieties. Consequently, there is room for raising the light use efficiency of crop plants by three to five times in the future.

3. Plant gene resources are the basis of crop breeding. Sustainable agriculture development necessitates the efficient collection and protection of plant germplasm

Human activities are exerting more and more pressure on the biological resources in China, resulting in the loss of biodiversity and fragmentation and degeneration of natural ecological environments in many regions. The next 20 years are key to the development of biotechnology industry in China. The regions with rich biological diversity and resources are strategic to this development. Thorough investigation and sustainable utilization of the existing biological resources in these regions are undoubtedly of paramount importance. During the 1950s to 1970s, China conducted many nationwide and regional inspections of biological resources. But in general we still know very little about the biological resources in many key regions. Moreover, past investigations focused largely on the identification of new species. Little attention was paid to species communities and the relationships among different communities. A full understanding of the characteristics of the biological resources has yet to be achieved for many important regions. Thus, it is urgently required that the biological resources in China are investigated, collected and evaluated, which should provide rich gene resources for modern crop breeding.

3.1.2 Analysis of Developing Trends

The main developing trends in the research on plant germplasm resources and crop breeding technology include the following aspects.

1. Breakthroughs in photosynthesis research will lead to revolution in crop breeding

A better understanding of the genetic control of carbon cycle in higher plants will provide new approaches for increasing the efficiency of photosynthesis[1]. Carbon assimilation, which requires the function of the key enzyme ribulose 1,5-bisphosphate carboxylase/oxygenase (Rubisco), is one of the major limiting steps for raising the efficiency of photosynthesis under natural conditions[2]. Genetically modifying Rubisco structure with higher specificity factor potentially improves photosynthesis efficiency[3], which will provide new biotechnology and bring bright perspective in improve crop breeding[4]. Compared to C3 plants, C4 plants have higher photosynthesis efficiency, lower compensation point of carbon dioxide, and low level of photorespiration. Under stress conditions such as high light, high temperature and drought, C4 plants generally exhibit better growth performance, and have higher water and nutrient use efficiencies. Moreover, C4 plants generally have higher biomass. The higher photosynthesis efficiency of these plants is associated with not only the special anatomy structure of their leaves but also their PEPC enzyme that has higher affinity to CO_2 than Rubisco and does not suffer from inhibition by O_2. For a long time, scientists have tried to build C4 CO_2 concentrating mechanism in C3 plants, increasing the CO_2/O_2 ratio of Rubisco, hence improving the carboxylation efficiency of this enzyme[5].

Studying photosynthesis mechanism will provide theoretical knowledge and potential technology to improve solar energy usage efficiency, further increasing crop yield potential. The possible limiting steps in solar energy usage efficiency of higher plants have been analyzed, and scientists are trying to genetically engineer these steps to increase photosynthesis efficiency. It is expected that solar energy conversion mechanism and regulation of photosynthesis will be thoroughly studied in the next 20 years, which will bring breakthrough in biotechnology for breeding new crop varieties with higher yields. Moreover, after harvest, the stalk and leaves can be used for biofuel production. For example, if photosynthesis efficiency increases three to five folds in the next 30 to 50 years, the biomass of some crops such as sorghum will be 30 to 60 tons per Chinese mu, which equals approximately 6 to10 tons of fuel ethanol per Chinese mu.

In future, at the same time of substantially increasing food production, efforts must also be made to invent new products to meet new consumer needs. These products include more nutritious functional foods, biomaterials with added medicinal value, and biomass for energy production. In the next ten years (2010-2020), "designer crops" will be tested, and their leading role in agriculture will be confirmed. The intermediate and long term targets include the development of new industries and new economy based on the efficient exploitation of plant biomass.

In C3 plants, the first step in the dark reaction of photosynthesis involves the fixation of CO_2 by the enzyme Rubisco into 2 molecules of 3-carbon compounds (3-phosphoglycerate). In C4 plants, CO_2 is incorporated into the 3-carbon phosphoenolpyruvate to form a 4-carbon organic acid (oxaloacetate).

C4 plants are distributed mainly in the tropics. Compared to C3 plants, C4 plants have higher CO_2 assimilation rate and photosynthesis efficiency, resulting higher biomass and grain yield.

2. Production of raw materials for bioenergy will become an important component of future macro-agriculture

The worldwide shortages of fossil energy usher in the era of biofuel industry. But the right to have sufficient food to eat supersedes the convenience offered by cars. Therefore, the development of future biofuel industry must rely on the use non-arable land and specialized energy crops. In China, because of the huge population and inadequate arable land resources, the cultivation of energy crops must not compete with that of food crops, or at the expense of decreased food security. The rational use of marginal land resources, such as saline soils, in combination of ecology conservation and rebuilding, should be a viable strategy for ensuring the coordinated and sustainable development of society, energy and ecology in China. The research on energy crops has just begun. Substantial breeding efforts are still needed to make available improved energy crop varieties. The progress of genomics will speed up the efficiency of breeding energy crops. Many improved energy crop varieties will be released in the next ten years. In the next 30 to 50 years, energy plants will become the most important industrial raw materials on earth.

3. Systems biology will provide theoretical and technological foundation for large scale of mining and utilization of gene resources

The main objective of systems biology is to systematically study the genes and their interacting networks involved in the control of one or more complex biological processes at the transcriptional (transcriptomics), protein (proteomics), and metabolic (metabolomics) levels, with the deployment of appropriate and high throughput technologies[6]. In transcriptomics research, the use of microarray hybridization analysis can reveal the transcriptional responses of all genes of a plant species to a given environment stimulus. This, together with further investigations using mutants, will lead to the understanding of the biological function of many genes in a relatively short period. In proteomics research, the deployment of high throughput separation and identification methods enables researchers to follow the changes of hundreds and thousands proteins under one or more conditions. For those targets showing significant changes, their amino acid sequences can be obtained with mass spectroscopic tools, which can be employed to further track down the coding genes of the target proteins. In metabolomics research, the deployment of mass spectrometry and nuclear resonance methods makes it possible to observe the changes of thousands of cellular metabolites caused by functional alterations of one or more genes, thus connecting gene function and metabolism. Moreover, the progress of transgenic technology will further simplify the generation of genetic mutants for more and more plant species. Consequently, it is clear that the advancement of systems biology will certainly lay down a solid foundation for large scale gene functional analysis, resulting in the isolation, identification and functional verification of hundreds and thousands of important genes. The flourish of the information on gene function will lead to a significantly better understanding of the molecular mechanisms underlying complex traits, such as the agronomic characteristics of crops. The knowledge and resources accumulated will eventually enable the design and improvement of the biological traits and species characteristics of crop plants.

4. Breeding by molecular design will produce novel crop varieties and eventually intelligent plants

The prerequisite for breeding by design is thorough understanding of the function of the genes or quantitative trait loci (QTL) controlling important agronomic traits. The gene functional insight will facilitate the development of novel germplasm lines, which can be regarded as individual modules required for molecular design breeding, by marker assisted selection (MAS), TILLING (Targeting Induced Local Lesions In Genomes) and/or transgenic approaches. The different modules are coordinately assembled together to form a variety according to the desired breeding goals. In this way, breeding by design provides not only an effective solution for dissecting and improving complex traits but also a highly efficient paradigm for crop breeding.

The methods for MAS of qualitative traits in plants are now well developed, and have been successfully deployed in crop improvement. There have also been encouraging achievements in MAS for quantitative traits. The development of the markers closely linked to important QTL permits the simultaneous tagging, transfer and pyramiding of the genes controlling quantitative traits, and thus the efficacy of using minor genes in crop breeding. For example, American scientists have succeeded in transferring and pyramiding eight QTL controlling sugar content in tomato fruits by MAS, which is unlikely to be achieved by conventional breeding. A huge advantage of MAS is that markers are used to follow gene and trait transfer in early generations, thus reducing the need for laborious phenotype identification and the time required for variety development[7]. Moreover, MAS breeding avoids the contention of bio-safety associated with transgenic breeding, and therefore has the potential to be further optimized as the safest and most practical plant breeding tool in the future.

The gene resources produced by functional genomics and systems biology research provide both the materials and impetus for transgenic crop breeding. Since the first field release of transgenic plants in 1986, many thousand cases of transgenic plants, which encompass more than 40 different plant species, have been put to field testing in 30 countries[8]. The cultivation of transgenic crops has also produced huge social and economical benefits[8]. Transgenic breeding not only enables cross-kingdom gene transfer, but also avoids genetic drag commonly encountered in conventional breeding. In addition, transgenic breeding is also less time consuming and more accurate in phenotype characterization when compared to conventional breeding. Thus, although there still exist disputes on the environmental- and bio-safety of transgenic plants, transgenic breeding will undergo further refinement, and is likely to become a key component of crop genetic improvement in the future.

Superior alleles created by artificial mutagenesis and elite QTL transferred from the wild relatives of crop plants expand the genetic resources for molecular breeding. TILLING is an effective approach for creating and identifying mutant alleles of a given target gene[9,10]. The series of alleles produced by TILLING project is not only useful for deeper understanding of the structural and functional relationship of the target gene, but also increases the chance of finding one or more superior alleles that are more useful in trait improvement than the native gene. Consequently, TILLING will play an important part, and has a bright future, in plant functional genomics research and crop molecular breeding in the years to come. Eco-TILLING, a further development of TILLING, is highly useful for mining the beneficial alleles of key genes in natural populations of crop plants[11,12].

The preparation of chromosomal introgression lines is in essence chromosomal engineering at the molecular level, which makes the use of alien genetic information in crop improvement into a more systematic and practical endeavor[13]. Considering that MAS is easily conducted in the main steps of the development and characterization of chromosomal introgression lines, and that the majority of the cultivated crops has closely related wild relatives or geographically distant ecotypes, it is certain that these lines will have significant impact on broadening the genetic basis and the creation of revolutionary germplasm materials and varieties of crops. Consequently, chromosomal introgression lines will be essential players in future crop breeding.

The key points of breeding by design, as originally proposed by Peleman and van der Voort[14], include obtaining a sound knowledge of the function of the genes and QTL controlling agronomic traits, creation of novel germplasm lines by MAS, TILLING and transgenic methods based on gene function, selection of appropriate germplasm lines according to specific breeding goals, and pyramiding multiple desirable traits by appropriately crossing the selected lines. Because gene and QTL transfer is monitored molecularly, and can be accelerated with transgenic technology when necessary, breeding by design has unmatched advantages in terms of time saving and high accuracy in gene and phenotype characterization. Although the concept of breeding by design was proposed only a few years ago, it is now widely accepted that this approach will become a formidable force in turning the results of functional genomics and systems biology research into the real progress in the breeding of elite crop varieties, and by doing so, creating huge economic benefits. The advent of the era of breeding by design will certainly revolutionize conventional breeding practice, and bring valuable opportunity for the much needed breakthrough in crop genetic improvement.

During the transition from conventional breeding to breeding by molecular design, the following key developments may be anticipated. ①By 2020, the technologies for efficient gene transfer and elite germplasm development become sophisticated in major food and oil crops. The integration of MAS and safe transgenic breeding enables effective transfer and pyramiding of key genes and their interacting networks. The varieties developed via molecular design will overcome the major limitations of the cultivars currently available, and become the major players in agricultural production. ② By 2030, as the technology and efficacy of molecular crop design undergoes further refinement, the molecularly designed varieties will replace conventional cultivars and gain wide applications, laying the foundation for sustainable development of agriculture and sufficient food supply. In the mean time, the plants and crop varieties molecularly designed for special purposes will begin to provide non-food resources for human consumption, such as clean energy. ③ From 2040 to 2050, intelligent plant varieties may appear as a consequence of further progress in molecular design breeding. These plants can automatically sense and distinguish various changes in the environment, and adjust their metabolism accordingly in an active, timely and quantitative manner so as to maintain optimized growth performance and yield the products suitable for human consumption while protecting the ecological environment.

3.2 Scientific and Technological Goals

3.2.1 Overall Goals

The overall goals include ① major breakthrough in plant photosynthesis and light use efficiency, ② development of genomics based key biotechnologies, and ③ construction of new technology system for breeding agricultural crops (including plants with specially designed function). To reach the above targets, approaches based on systems biology research will be taken, and the abundant gene resources in China and the world will be harnessed for the common good. It is expected that the realization of these targets will provide sound scientific and technological bases for sustainable agricultural development in China.

3.2.2 Important Subject Areas and Development Goals

To realize the overall targets listed above, innovative research is absolutely necessary in the following four subject areas: basic research and key enabling technologies, food production and security, bioenergy, and sustainable utilization of gene resources.

1. Basic research and key enabling technologies

(1) The molecular mechanisms of important agronomic traits

By studying the molecular genetic basis of agronomic traits, key genes and their functional interacting networks will be identified. By using the genomics and systems biology approach, the genetic relationships among the various gene networks determining different traits will be established. These efforts will yield both theoretical support and gene resources for molecular design breeding.

(2) Research and development of efficient and high throughput molecular chromosome engineering technology

Using crop relatives as elite gene donors and in combination with high throughput MAS, systematic chromosomal introgression lines suitable for individual agroecozones will be developed. These materials not only expand the genetic basis of crops, but may also directly serve as key components for molecular design breeding.

(3) Research and development of safe, efficient and high throughput transgenic technology

Based on the success of the transgenic technology in model species (*Arabidopsis thaliana*, rice), major breakthrough in maker free transgenic technology (MFTT) will be made for complex crop species (such as wheat, maize, soybean, cotton, etc). With further refinement, MFTT will become a safe, efficient and high throughput transgenic technology, allowing simultaneous transfer of multiple genes and tight spatial and temporal controls of transgene expression. The elite transgenic lines will serve as useful materials for molecular design breeding.

(4) Construction of basic infrastructure for molecular design breeding

As the research on molecular design deepens, crop breeding will become a reproducible science with sound scientific design and strong predictive power. This raises higher requirement on the basic infrastructure used for conducting breeding research. The integration of indoor agricultural facilities, biochemical and molecular screening platforms, and automatic control and information technologies will enable the establishment of a highly accurate infrastructure, which provides reliable support for the effective improvement of various traits (i.e., processing and nutritional qualities, pest resistance, resource efficiency, etc.) and molecular design breeding.

2. Food production and security

(1) Improving quality traits and raising yield potential and stability of crops

By applying molecular design breeding, and through coordinately pyramiding the key genes and QTL controlling yield (i.e., the numbers of tiller, spike and spikelets, thousand grain weight, fertility, etc), quality (i.e., starch, protein, fatty acids, etc), stress resistance or tolerance (i.e., pest resistance, dough and low temperature tolerance, etc), novel varieties with ideal architecture, high light energy use efficiency, superior quality, durable and wide spectrum stress tolerance will be developed. The strong negative correlation between yield and quality traits will be broken. The target of raising the crop yield potential per unit area will be realized.

(2) Developing novel varieties with salinity tolerance and high nutrient use efficiency

By mining and cloning the genes and QTL controlling salinity tolerance or nutrient use efficiency, and through transgenic and other molecular breeding methods, novel varieties that can be cultivated in saline or nutrient poor soils with high yield potential will be developed. These varieties permit the exploitation of saline and nutrient poor soils, alleviate the high input of chemical fertilizers in agricultural production and resultant environmental pollution, and largely increase the output of the arable land currently giving rise to only middle and low yield potential.

(3) Developing novel varieties tolerant to prolonged storage

The safe and long term storage of rice, wheat and maize grains is a strategic need of food security and social stability in China. But currently grain storage is often accompanied by rapid decreases of processing and nutrient quality, and in worst cases, by the occurrence and infestation of pathogenic fungi and their toxins, rendering the stored grains no longer suitable for food uses. Through elucidating the molecular mechanisms underlying the deterioration of stored grains and the processes of fungal infection and toxin production in the grains, novel varieties with strong tolerance to the adverse changes during storage will be developed, providing support for long term grain storage and food security.

(4) Developing health promoting, functional plants

The requirement of health promoting foods by consumers (particularly those suffering from specialized diseases) necessitates the research and development of functional plants. Through transgenic and other molecular breeding methods, the contents of the compounds beneficial to human health (such as vitamins, carotenoids, flavonoids and related polyphenols, etc.) in harvested crops will be increased to desirable levels, thus leading to the development and cultivation of functional plant varieties.

3. Bioenergy

(1) Mining bioenergy plant resources

Worldwide energy crisis and deterioration of global ecological environments propel the development of renewable energy resources. As a form of clean and renewable energy, bioenergy has received high attention from many countries. One of the bottlenecks limiting bioenergy progress is the insufficient supply of bioenergy materials. The arable land resources in China are limited. The growth of bioenergy must not be at the expense of food security. The cultivation of bioenergy plants must not compete with that of food crops, and be carried out using marginal land resources. Consequently, efforts must be made to identify potential bioenergy plants with high photosynthesis and biomass and strong tolerance to adverse conditions, and to improve the biomass accumulation per unit area of these plants in arid, saline or nutrient poor soils. These plants will provide a strategic basis for bioenergy development in China, and aid the breeding of new and environmentally friendly bioenergy varieties with even higher and more stable biomass yield in the future.

(2) Optimization of the cultivation of bioenergy plants

China has rich climate and plant resources. However, the present efforts in developing bioenergy are scattered, leading to high costs in biomaterial collection and transportation. The available research shows that sweet sorghum, cassava, sugarcane, *Jatropha curcas* are the energy plants with the highest industrial prospects. Cassava, sugarcane and *Jatropha curcas* are mainly cultivated under tropical and subtropical conditions. But in China the marginal land resources are very limited in tropical and subtropical areas, but largely distributed in the northern, northwestern and northeastern regions. Thus, optimization of the cultivation of different species of bioenergy plants according to the climate, ecological and economic features of different geographic regions is vital for the efficient development of bioenergy industry.

4. Sustainable utilization of gene resources

(1) Investigation and registration of plant communities and germplasm materials in strategic bioresources regions

Strategic bioresources have huge commercial values, and are key to national competitiveness in international economy. These resources, particularly those having new industrial prospects and crucial to solving the energy and environmental problems in China, will become an important component of the national economy. China is one of the distribution centers of strategic bioresources. But our investigation and understanding of these resources are still limited. Thus, it is urgent that the investigation, registration and rational exploitation of plant communities and germplasm materials in the strategic bioresource regions of China are conducted. In the mean time, new efforts should be invested to establish trait improvement systems of bioresource plants through artificial hybridization and transgenic breeding, and *in situ* protection areas or experimental farms for the multiplication and propagation of exotic accessions. Planning for industrial exploitation of bioresource plants should be developed based on proper assessments of their risks and benefits to ecological, social and economic developments.

(2) Development and creation of functional plants grown under special ecological niche

Plants grown under special ecological niche are those with high adaptability to extreme environmental conditions, the study of which can offer precious information on the evolution of genetic and functional diversities in plants. These plants are important strategic resources ferociously chased by many countries. In China's vast and varied terrain, there exist many extreme environments showing different physical and chemical properties, which are often populated by diverse plant species. Efforts are needed to study the plants living in extreme environments, which contributes to understanding the rule and evolutionary mechanisms of biosphere, particularly the interactions between plants and environment. This knowledge is highly valuable to the development of new plant varieties with strong tolerance to drought, saline, heat or cold through molecular design breeding, thereby providing novel solutions to combat the problems caused by dwindling arable land and water resources.

(3) Development and creation of environmentally friendly plants

By understanding the mechanisms behind the metal tolerance of certain plant species, and through transgenic and molecular design breeding, the plants with large biomass, high adaptability and strong capacity to take up and accumulate heavy metals will be developed. These plants may provide a timely and effective solution to solving the heavy metal pollution problem in the soils. In the mean time, the knowledge on metal acquisition and accumulation by plants may aid the breeding of the food crop varieties with much reduced accumulation of toxic heavy metals, thus protecting the health of consumers.

(4) Development and creation of medicinal plants

China is among the countries that use abundant medicinal plants from very early ancient times. As the ecological conditions undergo deterioration, wild plant resources are on the decline. Efforts are needed to mine medicinal plant resources, dissect the biosynthetic and metabolic pathways of important chemical compounds, and develop novel cell lines and varieties with increased contents of bioactive ingredients through regulating appropriate gene and enzyme activities via molecular design breeding. The development and application of plants as bioreactors for medically useful compounds will contribute to the optimization of the infrastructure of agriculture.

(5) Development and creation of functional ornamental plants

Through collecting and mining germplasm resources, the genes controlling important ornamental commercial traits will be discovered. On this basis, and by combining conventional and molecular design breeding methods, novel ornamental plants with rich diversities in color, morphology, fragrance, and health promoting nutritional function will be created. The products from these plants will serve to improve the food consumption patterns, and to raise the mental and physical happiness and overall living quality, of human beings.

3.3 Roadmap for Scientific Development

3.3.1 Scientific Tasks

In future, scientific tasks for plant germplasm resource and molecular breeding will mainly include the following four aspects.

1) Evaluating their application potentials, and establishing proper evaluation system and developing means of sustainable utilization for strategic plant germplasm and gene resources by means of genome sequencing and bioinformatics methods (barcoding DNA, identifying and isolating useful genes and analyzing their functions).

2) Building platforms for "omics" (genomics, proteomic and metabomics) and systemic biology researches for large scale plant gene mining, molecular design, and molecular improvement and molecular screenings of plant varieties.

3) Establishing and completing the concept for molecular assisted breeding by design and its technical system for strategic plants in the country.

4) On the bases of identifying the key genes their networks that control important plant traits, developing plant varieties with high yield, functionally versatile and quick responding to environmental changes, as well as new varieties with superb bioenergy traits, e.g. massive biomass, highly resistance to stresses, and so on.

1. Establishing evaluation criteria and means of sustainable utilization for germplasm and gene resources

(1) Survey of plant community and plant germplasm resources for potential strategic bioenergy bases

The survey will be carried out on three levels: community, population and species and the contents should include: ① To conduct ecological community survey based on specific areas and sample plots in the zone with territorial control, and draw the distribution maps of ecological community categories and bio-resources; ② To collect the critical resource and the germplasm resources of key species, including plant seeds, unorganized materials and dipped specimens etc; and ③ To establish the database and user sharing system for ecological community and germplasm resources.

(2) Survey of wild relatives of some strategic bioresources, their conservation and development planning

It will be carried out by combining genome sequencing and bioinformatics methods to bar-coding DNA, identifying and isolating useful genes and analyzing their functions; the economical value of surveyed plant resources will be evaluated comprehensively and their important traits that have potentials for marketing will be considered for enterprising planning on the basis of consultation on benefits as well as risks for ecosystem, society and economy. The final goal would be to put forward some concrete enterprising development planning to scientifically identify and utilize rich plant resources in China through close cooperation among national research institutes and enterprising planning sectors.

(3) Exploring plant resources for bioenergy

The germplasm database of bioenergy plant species should be established on the basis of integration and assortment of currently available bioenergy plant resources. Subsequently, the core germplasm comprising important bioenergy plant species will be established and genetic diversity will be analyzed, emphasizing on the assessment of their bioenergy related traits over different habitats or ecosystems. The cultivation methods for high yield production should also be studied systemically across arid, semi-arid and saline land. The storage physiology for bioenergy plants should be studied and their effective storage measures be formularized. To meet the goal of breeding varieties with high biomass, good stress-resistant and excellent energy related traits, platforms should be built up for molecular breeding related researches and practices, such as molecular design, molecular improvement and selections, networks of metabolism and their regulations, stress-related traits selection and improvement, as well as an assessment system for effect on environment and biodiversity in case of large-scale bioenergy cultivation.

(4) Structuring bioenergy plant cultivation

It is important to optimize the cultivation pattern across the country according to the ecological and economical features of different regions. For example, in northwest China it should be mainly for planting sweet sorghum and arid shrubs, while in southwest and south China it should be mainly for planting woody oil plants, cassava and sugarcane, and in the non-cropland in northeast and north China it should be mainly for production of bioenergy plants like sweet sorghum.

2. Completing the concept for molecular design breeding and its technical system

(1) Elucidating molecular mechanisms underlying key crop agronomical traits

A prerequisite to crop trait and variety designing is the understanding of molecular mechanisms underlying agronomical key traits of this crop. Future works should emphasize on: ① studying key genes and their networks that regulating key agronomical traits using rice as a model crop; ② carrying out comparative biology studies among model plants (e.g. *Arabidopsis*, rice) and complicated crops (e.g. wheat, maize, sweet sorghum, soybean, rapeseed). It will help to identify key genes and up/down stream components and their interactions, and interactions between gene networks behind related traits, most likely to obtain the key gene resources with high economical value. Further effort should put on establishing an interconnected system (or association system) that will include key genes controlling high yield, good quality, stress resistance and high efficient utilization for water, light and soil nutrient elements, which will provide theoretical basis as well as gene resources for making breakthrough in molecular design breeding.

(2) Developing high efficient and large scale molecular chromosome engineering techniques

With the establishment of high-throughput molecular marker techniques (specially, the SNP identification and detection techniques) molecular chromosome engineering (e.g. systemically creating introgression lines in a large scale) will be developed with better opportunity, and the chromosome introgression lines should play more and more important roles in expending crop genetic diversity. It is necessary to consider the following aspects in the area of chromosome engineering: ① for major crops like rice, wheat, soybean and cotton, systemically creating and accumulating chromosome introgression lines using suitable cultivars as recipients and proper related species as donors; ② establishing convenient and accurate genotyping system to define the background of the introgression line, the similarity with the recipient and insertion information; ③ discovering and accumulating elite introgression lines based on evaluation of their agronomical traits in certain suitable ecological regions, improving promising candidates as components for molecular trait and variety designing.

(3) Creating large scale, safe and efficient plant transgenic technology through R&D

Unlike the situation for few model plants (rice and *Arabidopsis*), it is in urgent need to further develop and optimize stable transformation techniques emphasizing on techniques such as marker-free, efficient and stable expression, temporal and spatial regulated expression and multiple gene transformation. In the mean time, platforms building will be inevitable for the application of transgenic technique on crop improvement. Those platforms must be standardized with reasonable scale. To do so, necessary studies should be focus on following things: firstly, development of techniques, such as maker-free selection, efficient and stable expression, temporal and spatial regulated expression and multiple gene transformation, complicated and vital crop and plant resources; secondly, standardized platform building-up for large scale phenotyping in those targeted major plant cultivation areas, trait assessment for transgenic lines, discovering of elite transgenic and accumulation of molecular design components. The purpose is to combine traditional breeding methods with transgenic approaches to better serve practical crop improvement.

(4) Building up facilities for molecular design breeding researches

The crop molecular design breeding will gradually become a practice so that results can be designed, reproduced and predicted, thus it is necessary to have basic research facilities built-up to meet those requirements. It is specially the case when the consumption becomes more and more individualized and requires facilities for trait measurement and integration. Those facilities should serve the following purposes: ① testing and evaluating specialized grain quality; ② testing and evaluating disease and pest resistance level; ③ testing and evaluating tolerance and resistance to high and low temperature; ④ water-saving variety selection and breeding; ⑤ light using efficient and fertilizer saving variety selection and breeding; ⑥ salt tolerant variety selection and breeding; ⑦ bioenergy plant variety selection and breeding.

3. Breeding high yield and good quality new crop varieties and improving crop traits that help to stabilize yield

(1) Improving crop traits related to yield stabilization and maximizing yield potential

It is believed that the traditional breeding approaches are reaching the bottleneck for yield increasing per unit and for resolving the conflicting between quality and yield in short period of time.

The emerging and the application of molecular design breeding technology should hopefully provide means for solving the above difficulties. Crop yield is a complex traits, comprising planting density, branching, spike number, spikelet number, grain weight, fertility and harvest index, which are QTL controlled. Large number of yield-related QTL have been located by Genetists and breeders, however, their usefulness in breeding application still has a long way to go. It is very important to continuing fine mapping, cloning and using those QTLs for the purpose of speeding up genetic improvement of multiple traits and shortening breeding cycles. Biotic and abiotic stresses are the major factors that affect crop yield potential. These stresses include pest diseases, drought and low temperature. The goal of future breeding research and practice should be the integration of multiple genes targeting disease and insect resistance, drought and low temperature for creating new varieties with durable and broad spectrum resistance and stress tolerance, through combination of traditional breeding with modern biotechnology breeding approaches.

One way to maximize rice yield potential and to make a breakthrough for rice yield should be to combine heterosis with ideal plant architecture. The advantage of this combination is to increase the lower blade light use efficiency while reducing the pest and disease occurrence in the canopy.

(2) Improving crop grain quality traits

Grain quality is a comprehensive trait, including nutrition quality, processing quality, and food palatability. For example, in rice grain the amylose content and the structure and feature of amylopectin in rice grain determine its cooking and processing quality as well as the food palatability, whilst wheat grains contain hundred of glutenin, the content and the combination of different glutenin subunits determine wheat grain processing quality. Study of the interaction of multiple genes and the effect on grain quality, and combining several genes affecting quality in one cultivar through biotechnology, will certainly help to improve a cultivar's comprehensive quality trait.

(3) Creating new varieties that are tolerant to salinity and barren land while maintaining high nutrient use efficiency

It is necessary to strengthen researches on crop salinity tolerance and to clone genes from wild related species and organisms living in extreme conditions so that to enlarge the gene pool of salinity tolerance. New varieties, developed from combined effort of traditional and biotechnological breeding approaches, will grow on soils with high salt content to buffer the sloped crop productivity in intensively cultivated area where large amount of fertilizer has been applied and payoff of fertilizer is reduced, and where there are increased concerns about environmental pollutions due to fertilization.

4. Breeding new varieties with specific exceptional characteristics through R&D

(1) New varieties suitable for longer grain storage

Storage of staple crop grains is important for food safety and stabilization of the Chinese society. However, it is common that during storage the grain quality and its nutrition value decrease rapidly whereas toxins content may increase due to infections from pathogens like *Fusarium*, *Aspergillus flavus*. The commercial value will be lost completely in some cases in few years and thus huge losses for the country. It is inevitable to carry out researches focusing on grain storage, for example, the molecular mechanisms behind rapid nutritional and quality reduction of staple crop grains in certain storage conditions, the molecular mechanisms controlling grains infection by pathogen and thus toxin accumulation, the effect of various biological agents on grain storage, as well as new grain storage technology development.

(2) New functional crop varieties beneficial for health

The demand for health foods make it necessary to study and develop new functional crop products that benefit human health and fit particular groups. Some crops contain active substances with health promotion functions, for example, antioxidants (vitamins, carotenoids, flavonoids, and polyphenols, etc.) for preventing ageing related diseases, beneficial cholesterol (HDL), unsaturated long-chain fatty acids for reducing cardiovascular disease rate, and bioactive peptide for preventing or curing hypertension. However, the content of these components is usually below the functional threshold in natural plants. Thus, it is very important to increase their expression level or to induce the production of bioactive ingredients in plants through transgenic and breeding strategies.

(3) New varieties with medical and medicinal functions

China is one of the countries that utilize natural plants for medicinal purposes in the earliest time and in the largest amounts. With the shortage of related wild plant resources, it is urgent to develop new approaches for the production of active substances that can be used for therapy. One approach is the development of plant bioreactors to produce substances like glucagons for diabetes treatment, vaccines and antibodies for curing infectious diseases of animal origin. Therefore, cell lines optimized through biotechnology for producing high content of those bioactive substances, plant tissue culture system that can expanded into industrial scale will also be necessary for commercialized production in the future. Large scale cell culture and industrial scale production should be one of the promising interdisciplinary technologies for obtaining a variety of medicines. The reasons are, firstly this technology does not depend on the land availability, and secondly, on the basis of understanding of their metabolism and the biochemistry of certain secondary metabolites it should be possible to regulate the expression of related genes and synthesis of their products in this system.

(4) New functional crop varieties for special habitats

Many regions in China are not suitable for growing crops because of water resource shortage and climate conditions. The situation is getting even worse due to the increasing frequency of natural disasters and extreme weather conditions, such as draught, high temperature and freezing damage. The consequences of these are the ever worsening situation for the shortage of arable land and water resources, as well as huge losses in agriculture production resulted from abiotic stresses.

Researches on the mechanisms of stress resistance or tolerance and identification of related key genes brought about some findings that can applied to crops, resource plants and environment-protecting plant varieties that grow in different regions and climate conditions in the country. These findings will be valuable for agriculture productivity, environment protecting and repairing and so on. Plants growing in extreme conditions have developed superior capability to adapt to the surroundings as the results of long time adaptation, therefore, these plants represent a wealth of genetic/functional variations and information accumulated from their evolution. Studying these plant species should be extremely important for revealing the law of biosphere development, the mechanism of biodiversity formation and evolution, and the interaction between life and environments. There are ideal conditions to carry out studies on functional crops living in special habitats in China because so many plant species living in those places in this vast land, examples including high mountain *Chorispora bungeana* resistant to freezing temperature, *Puccinellia tenuiflora* tolerant to salinity, and the resurrection plant *Boea hygrometrica* surviving extreme dehydration, and so on.

(5) Crop varieties that are environment friendly

One of the problems urgently need to be addressed is heavy metal contamination in China. Phytoremediation is one of the ideal approaches for clean these contaminations in soil. Compared with traditional chemical and physical approaches, phytoremediation takes advantage of plant resources from nature or developed from genetic breeding methods so it has the advantages of simple, economic and without secondary pollution. In-depth Studies of molecular mechanisms of plant heavy metal enrichment are important guarantee for agriculture production in China and the exportation of Chinese agricultural products to world markets. In the mean time, a system should be established to collect plant resources suitable for metal enrichment and to develop effective ways for utilizing them.

(6) Functional horticultural plants

Horticultural plants are important for human because they are rich in carbon nutrients, organic acids, amino acids and proteins, and lipids, which can be used for improve human diet, while the color richness and the forms and fragrant of horticultural plants can bring spiritual and physical pleasure to human. Studies revealed some horticultural plants contain substances that are good for human health, and even for curing diseases. Future studies should focus on important functional horticultural plants, including fruit trees (for example, grape, peach, kiwi and blueberry) and flowers (for example, peony, lilac, lotus and lily). Major aspects should be considered, such as collection of germplasm, evaluation of important quality and resistance traits, isolation and analysis of their functional components, assessment of genetic variation, core germplasm establishment, identification and isolation of key genes. New germplasm and breeding materials will be developed through traditional hybridization breeding combined with modern molecular design breeding.

3.3.2 The Roadmap

1. Roadmap rationale

It is the fundamental society needs that drive the completion of this roadmap. The needs include developing novel crop varieties that are multi-functional, high yield and high quality, some of these varieties are looked as intelligent, in that they should be able to responding to environmental changes and to activating or silencing quantitatively gene expressions and translations. The needs also include to assuring food safety for Chinese society, elevating life quality and individual health level, and meeting the industrial raw material needs for biofuels, and so on. The Roadmap starts from the fully uncovering and utilization of gene resources that are rich for some plants or characters, addressing the regulation of key genes and their associated networks, and establishing and completing nationwide strategic conception framework of plant molecular design breeding, and in the mean time associated technological supporting systems. At a later stage, the gene resource stocks of strategic plants should be built on the bases of genomics understanding of most species and most important metabolic pathways, and enabling the scanning and assembling of ideal plant genotypes for the purpose of designing and realizing novel plant genotyping. The technical approaches for fulfilling the above goals should be integrating system biology, bioinformatics, functional genomics, and to interconnecting life science, information science and material science.

2. Roadmaps

Two maps have been designed for the plant germplasm resources and modern breeding areas: the science development roadmap (Fig. 3.1) and the timing diagram (Fig. 3.2).

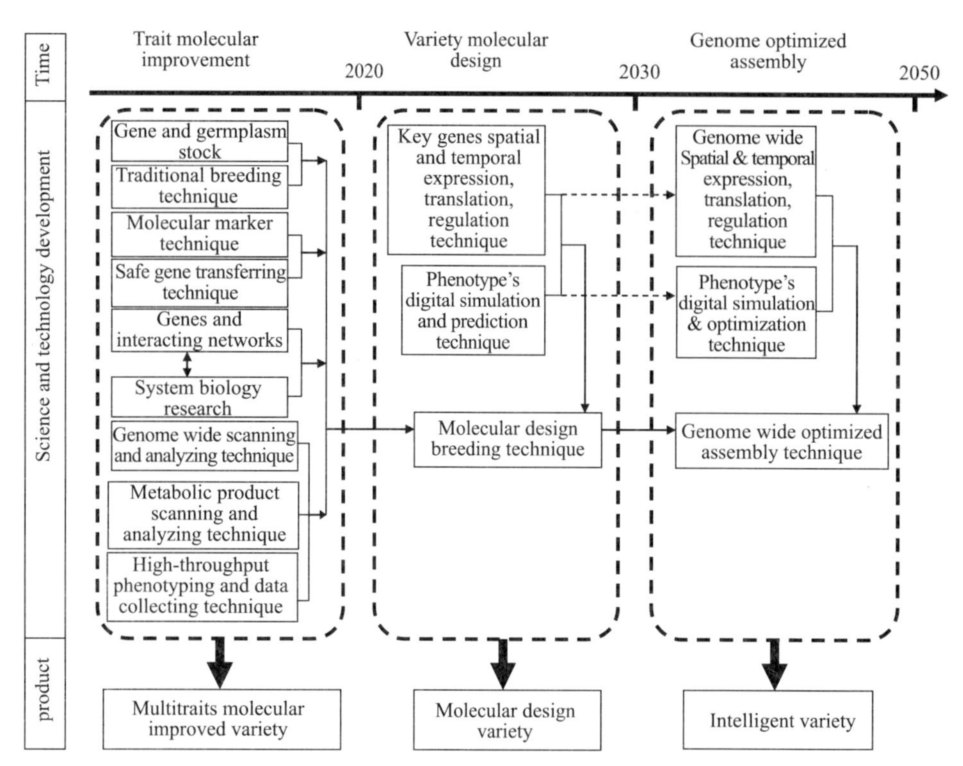

Fig. 3.1 Roadmap of plant germplasm resources and modern breeding areas

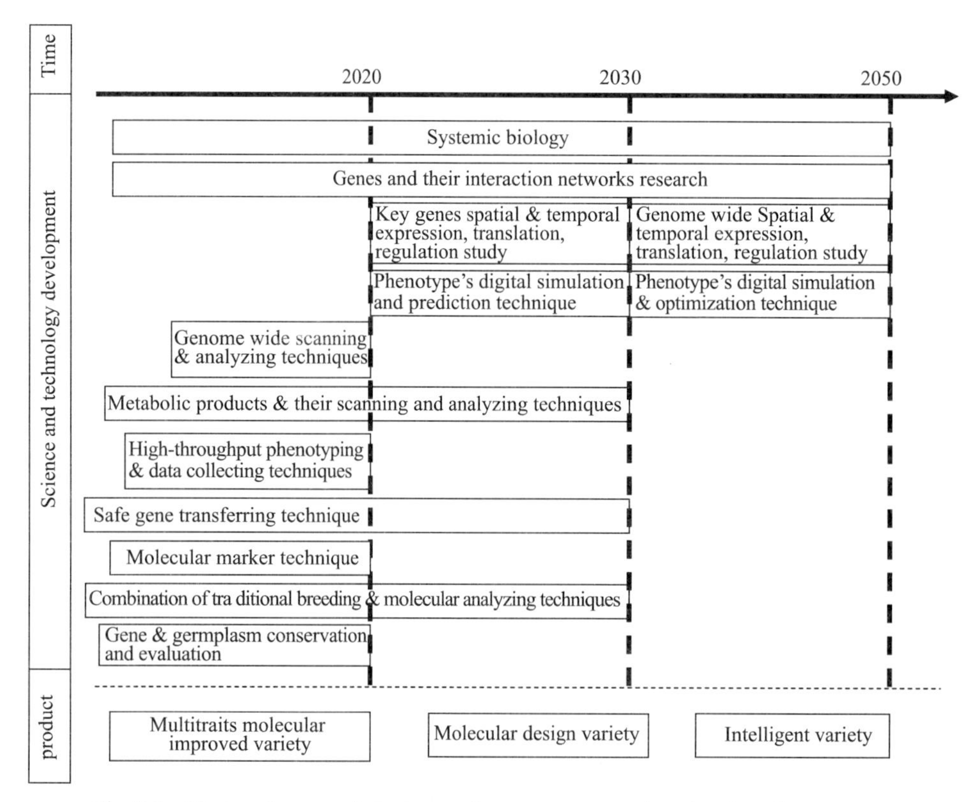

Fig. 3.2 Timing diagram of plant germplasm resources and modern breeding areas

(1) Evaluation and sustainable uses of plant germplasm and gene resources

Until 2020, the distribution maps for different ecosystem types and biology resources should be drafted based on large scale investigation, and the database platforms and data sharing systems should be constructed, which includes germplasm of key plant resources, seeds and explants, and so on. Until 2030, pedigree relationships, distribution maps and population dynamics of some representative resources should be finished, and strategies for protection of pivotal plant resources and their wild relatives, either by on-site conservation or botanic introduction conservation, should be finished. Until 2050, following tasks should be finished, i.e. DNA barcoding pivotal plant resources and germplasm and isolating and confirming the function of valuable genes; selecting the characters of most potentials in the market for sustainable utilization; Converting a group of important plant resources by molecular design breeding and channeling into industrial pipelines based on evaluation results for ecosystem, economic efficiency and risks; Elucidating plant stress mechanisms and using key genes in agriculture, which will be pivotal for the development of featured plant resources and environment protecting crops and therefore for the enhancement of agriculture efficiency and environmental conservation and repairing.

(2) The combination of tradition and modern breeding techniques breed varieties with multiple traits improved

Traditional breeding approaches have been very effective in agricultural practice by using hybridization, backcrossing, mutagenesis, and targeted selection for elite traits or combinations. However, the shortcomings of these approaches are the requirement for the long period of time for multiple traits improvement and for breaking unfavorable genetic linkage and reproductive isolation. On the other hand, the modern breeding techniques have advantages for overcoming those difficulties to achieve effectively multi-trait improvement based on the findings or methods of systemic biology.

Until 2020, safe transgenic technology should have been developed, alone with other biotechnology and new disciplines that allow high throughput data collecting and individual genetic constituents and their metabolic product scanning, together this should make it possible to simultaneously transfer several genes and improve multiple traits.

(3) Molecular design breeding and related technical system bring about novel crop varieties

Molecular design breeding should be able to assemble multiple genes according to preset goals using proper designing components derived from elite germplasms that are collected according to the knowledge of key genes or QTLs and guaranteed by technical supports like marker-assisted selection, TILLING techniques and safe transgenic technology. Technically, molecular design breeding will first locate loci controlling agronomical traits in a targeted manner, and then allelic variations of the loci will be surveyed in the germplasm pool, in the end the best combination of the loci will be assembled to fit the standard of elite lines or varieties.

Until 2020, new concept of 'designing crops' should be put forward and its verification be completed. The combination of MAS and safe transgenic technology should enable efficient transferring of major effect genes and their interacting networks in the major grain and oil crops. Until 2030, molecular design breeding technical system should be completed with the integration of other techniques, i.e. spatial and temporal gene expression, translation and modification, phenotyping digital simulation, therefore, the varieties developed should be spread over widely because most limitations of current varieties are overcome. Until 2050, molecular design breeding technical system should become the core breeding technical system and widely apply to the invention of intelligent plant varieties.

(4) Breakthroughs in gene network regulations and bioinformatics, alone with molecular design breeding, assist the invention of novel intelligent plant varieties

Whole genome scale gene expression, translation and modification should be possible after thorough studies on the regulation of key genes and their products. From 2020 to 2030, molecular design breeding should mainly target major grain crops, feed crops, muti-functional crops, plants with medicinal and health-promoting functions, and novel varieties of these plants or crops should be obtained.

From 2030 to 2050, a system consisting of molecular designing modules (i.e. traits, gene pathways and associated gene networking) and genome bioinformatics technology should be established, together with digital simulation and optimization technology that is derived from digital phenotyping simulation and prediction technology should be able to design and assemble intelligent plant varieties according to needs. These varieties could respond rapidly to changes of environmental key factors and optimize metabolic pathways in a timely and quantitatively manner, as such plants could maintain their best growth and developmental status to efficiently produce products that meet the human consumption as well as the environmental needs.

3. Routes for important fields

(1) Molecular design breeding targeting important agronomic traits and creating novel intelligent plant varieties

From 2020 to 2030, modern breeding technology centered around molecular marker assisted technology should be transformed from improvement of multiple traits to molecular design breeding. The major goals of molecular design target comprehensive improvement of multiple key traits, including yield, quality, nutrient use efficiency, pest and pathogen resistance and abiotic stress resistance traits. The content will be different for diverse crops and specific varieties that grow in distinct environmental settings. For example, the goals for wheat molecular breeding until 2030 include significant enhancement of gluten quality, increased nitrogen and phosphate use efficiency by 20-30%, increased water use efficiency by 10-20%, and in the mean time disease resistant against multiple pathogens like powdery mildew and rusts.

From 2030 to 2050, whole genome assembly based on genomic and bioinformatics technology should be conducted for the screening of elite genotypes and rapid optimized assembly according to needs, and intelligent plant varieties should be developed based on molecular design breeding technology and regulation on whole genome scale (whole genome assembly and gene expression, translation and modification, etc.), these varieties should maintain the best growth and developmental status by switching on or off gene expression networks, integrating relevant metabolic pathways for rapid responding to diverse environmental conditions. For example, the purpose of disease resistance of intelligent wheat varieties may no longer be immune to pathogens, but restricting or reducing the propagation of invading pathogens by rapid integration of disease resistance signaling pathways, or avoiding the invading of the majority of pathogens according to the pathogen race changing dynamics in population.

(2) Creating novel plant varieties of high light-use efficiency and super light-storage capacity

Theoretically, crop light use efficiency can reach up to 5%. Until 2020, it is possible to increase light use efficiency by 1% through selection of C4 plants and modification of structure of key Rubisco enzymes, and to establish energy plant germplasm and database for the selection of energy plant varieties.

Until 2030, major breakthrough should be possible in the mechanism and regulation of photosynthesis, and of metabolic pathways for plant energy converting and storing; in the mean time, molecular design targeting energy and stress-related traits should bring about super energy plant varieties.

Until 2050, light use efficiency should increase again by 4% and plant yield increase by 3-5 folds, biomass yield will reach 450-900 tons per hectare that is equivalent to 90-150 tons of ethanol per hectare. A large number of high energy plant varieties are developed and used in bioenergy industry.

Main References

[1] Kuang T Y. The principle and Regulation of Primary Light Energy Conversion Process during Photosynthesis. Nanjing: Jiangsu Science & technology Publishing House, 2003

[2] Spreitzer R J, Salvucci M E. Rubisco: structure, regulatory interactions, and possibilities for a better enzyme. Annu Rev Plant Biol, 2002, 53: 449-475.

[3] Ashida Y, Saito C, Kojima K, et al. A functional link between RuBisCO-like protein of Bacillus and photosynthetic RuBisCO. Science, 2003, 302: 286-290.

[4] Mann C C. Crop scientists seek a new revolution. Science, 1999, 183: 310-314.

[5] Matsuoka N C, Furbank R T, Takayama H. Molecular engineering of C4 photosynthesis. Annu Rev Plant Biol, 2001, 50: 333-359.

[6] Ideker T, Galitski T, Hood L. A new approach to decode life : systems biology. Annu Rev Genomics Hum Genet, 2001, 2: 343-372.

[7] Moose MP, Mumm RH. Molecular plant breeding as the foundation for 21st century crop improvement. Plant Physiol, 2008, 147: 969-977.

[8] James C. Global status of commercialized biotech/GM crops: 2008. ISAAA (International Service for the Acquisition of Agri-Biotech Applications, Ithaca, NY) Brief, 2008(39).

[9] McCallum CM, Comai L, Greene EA, et al. Targeting induced local lesions IN genomes (TILLING) for plant functional genomics. Plant Physiol, 2000, 123: 439-442.

[10] Slade AJ, Fuerstenberg SI, Loeffler D, et al. A reverse genetic, nontransgenic approach to wheat crop improvement by TILLING. Nat Biotechnol, 2005, 23: 75-81.

[11] Comai L, Young K, Till B J, et al. Efficient discovery of DNA polymorphisms in natural populations by Ecotilling. Plant J, 2004, 37: 778-786.

[12] Wang J, Sun J, Liu D, et al. Analysis of Pina and Pinb alleles in the micro-core collections of chinese wheat germplasm by Ecotilling and identification of a novel Pinb allele. J Cereal Sci, 2008, 48: 836-842.

[13] Zamir D. Improving plant breeding with exotic genetic libraries. Nat Rev Genet, 2001, 2: 983-989.

[14] Peleman JD, van der Voort JR. Breeding by design. Trends Plant Sci, 2003, 8: 330-334.

4 Roadmap of Animal Germplasm Resources and Modern Breeding Science and Technology Development

To 2050, with the doubling demand for livestock, poultry and fishery products, quality and safety of the products face serious challenges. The quantity and quality safety protection of livestock, poultry and fishery products and the restructuring promotion of agricultural production need significant breakthroughs in the research field of animal germplasm resources and modern breeding biotechnology. The scientific and technological advance can provide important technical support and guarantee for the development of modern livestock, poultry and fishery products.

4.1 The Requirements, Significances and Trends of Development

4.1.1 The Requirements and Significances Development

1. The increasing demand for animal protein will be a major national requirement, which can promote the development of animal germplasm resources and modern breeding biotechnology

Since the reform and opening-up, the rapid development of agriculture greatly improves the agricultural supply level; meanwhile, the rapid growth of the national economy, urbanization and development of food markets have rendered the structure of China's food consumption and dietary, residents, no matter rural or urban, have a substantial increase consumption in pigs, cattle, mutton, poultry and fishery animals, both of all contribute rapid changes in agricultural economic structure. However, compared with developed countries, the average per-capita amount of livestock, poultry and fishery products is still relatively low. Now, the average per-capita amount of livestock and poultry products just equal to the world level, however, compare with more than 100 kg in developed countries, the average per-capita amount of fishery products is only 30 kg, which remains a gap between 3-fold. In the level of dietary protein, China is still mainly with vegetable protein, animal protein is only 28% of the world average level, and 13% of developed countries such as the United States and France. With the development of China's economic and the improvement of people's living standards, consumer demand for livestock, poultry and fishery products and product quality will further increase, both of these will become the important driving forces for animal germplasm resources and modern breeding biotechnology.

2. The continue increasing demand for animal protein depends on sustainable advance in freshwater and marine fisheries

As early as 1995, for the first proposed "sustainable development", Lester R. Brown, was named "the world's 22 most influential economists," who was one of the famous American ecological economist and director of the Earth Policy Institute, he published a monographs called "Who will feed China?", and in that monographs, he prophesied that against the backdrop of the world's food production and fishing grew slowly or stagnated, as China increasingly serious water shortage, high-speed industrialization processed a large number of erosion on agricultural land, and population growth, in the early 21st century, China may have to import large quantities of grain from abroad in order to feed more than 10 million people, which may rise food price, and produce a huge impact to the world's food supply. After 14 years, China did not appear food crisis which was predicted by Lester R. Brown, Chinese people feed themselves. And other, compared to 37 countries which emerged food crisis, tilling 7% of the world's total cultivated land, China has nevertheless succeeded in feeding a population that makes up 22% of the world's total, what's more, with the rapidly development of aquaculture, China now supply world with high-quality and efficient animal protein—fishery products. In 2008, Lester R. Brown pointed out that, in the past 30 years, one of China's contributions to the world was fresh water fishing. He believes that "the well-development of China's fresh water fisheries is a great contribution to world", meanwhile, He said that, different from developed countries using the grain for feeding cattle and other livestock animal protein method, "fresh water fishing in China is extensive, this method is probably more efficient than that of developed countries, and can save the grain. This is the world the most efficient biotechnology"[1].

In the 21st century, marine is a valuable asset and the largest space for human society. Because of the world's population first and the relative shortage of land, fresh water and other strategic resources, marine plays a particularly important role for China. With 18,000 km of coastline, 3 million square km of sea areas under the jurisdiction of and over 6,500 coastal islands; China is rich in marine biological resources, such as fish, shrimp, shellfish, algae and other traditional fishing resources, what' more, may be rich in some rare biological resources of not more fully understand or impossible to land.

China is so densely populated, agricultural development is faced with the dual pressures of diminishing arable land and increasing population; per-capita cultivated land is only 1/3 of the world average, and more than 1,000 acres annually in recent years is still decreasing at a rate, so food safety security situation is grim. The bright future of efficient development of marine fisheries and beach will meet the growing demand for quality protein and make greater contribution. Since 1990, China's seafood production has been highest in the world, to reach 28 million tons in 2005; after 30 years of development, maricultural production has grown from 100 million tons in 2005 to 1,300 million tons in 2006, and the output value more than 200 billion Yuan. Marine fisheries have become one of the fastest growing and highest efficiency of agriculture in China, and provide effective protection for the demand of animal protein in China.

Thus, China's fishery development and fishery products' contribution to the world have been the concern of international economists, attracting world attention. Similarly, the output and quality of livestock, poultry and fishery products need the concept of sustainable development.

3. The scientific and technological advance can provide important theoretical support and technical guarantee for the yield and quality improvement of modern livestock, poultry and fishery products

To meet the increasing demand for livestock, poultry and fishery productions, the agricultural production needs restructuring promotion accordingly. The more demand of products, the higher proportion of animal husbandry and fishery in agriculture, at the same time, as the high value-added, labor-intensive agricultural products, livestock, poultry and fishery products in the international market have certain comparative advantages, the export of products have greater growth potential. In short, the development of animal husbandry and fisheries will be good for the farmers and people's living standards; they even play for China's agricultural comparative advantage, and have important significance to the sustainable and effective use of agricultural resources.

The scientific and technological advance can provide important theoretical support and technical guarantee for the yield and quality improvement of modern livestock, poultry and fishery products [2]. To improve the yield and quality of livestock, poultry and fishery products and develop new products with greater value, we must rely on science and biotechnology, and play weight on the increasing role of molecular design breeding and cell engineering breeding biotechnology. Many countries, especially developed countries, have taken conservation and utilization of fishery animal germplasm recourses and molecular design breeding biotechnology as one of the key knowledge-based economy in the 21st century. In Europe, many governments have predicted that, after 2010, more than 70% of agricultural products will depend on biotechnology. Therefore, in the world today, to enhance the conservation and utilization of fishery animal germplasm recourses, and to invest molecular design breeding and cell engineering breeding biotechnology, for many countries, have become important development strategies.

4.1.2 Trends

Since the 80s of 20th century, China has made some breakthrough in the basic and applied research field of livestock, poultry and fishery biotechnology, and generated a crop of key biotechnologies, formed a number of mutually complementary and practical value of biotechnology system, such as nuclear transfer biotechnology, sexual control biotechnology and transgenic biotechnology. The system development of these biotechnologies not only once had a lead over the rest of the world in basic research, but also attained clear results in the applied field, and bred a number of new varieties which contributed to industrial development and progress. To 2050, four mainly major trends in the field of fishery animal germplasm recourses and modern breeding biotechnology are as follows:

1. The genome research and genetic recourses study on livestock, poultry and fishery animals will promote new revolution in husbandry and fishery

Life sciences and biotechnology have made progress and development in recent years, particularly, the implementation of the Human Genome Project has shown business opportunities, and developed countries pay increasing attentions on fishery animal biotechnology, and have set up a number of researches and development companies accordingly, for example, the United States has implemented the "National Animal Genome Research Program, NAGRP", EC launched the "Pig Genome Mapping Project, PigMap and Bovine Genome Mapping Project, BovMap", Japan developed the "National Animal Genome Analysis Program, NAGAP", United States, Norway, Canada and other countries carried out the tilapia, rainbow trout and Atlantic salmon genome sequencing and functional genomics research. According to the U.S. Science Lifetime Achievement Award winner, Professor Soller and the European pig gene mapping project leader, Professor Haley predicted that, economic animal genome research will make the new changes in the field of animal husbandry, feed conversion rate of economic animal will make achieve to 2:1, annual milk production by cattle will increase to 13,000 kg, the daily gain of pigs will be up to 1200 grams, and lean meat percentage will be up to 75%.

2. Animal cloning biotechnology and transgenic animals will show a great potential

In aspect of animal cloning biotechnology, in early 1997, since the British scientists of Roslin Institute announced that they cloned the world's first somatic cell sheep, it once aroused worldwide attention, and was considered as a major breakthrough of animal biotechnology and life sciences, Governments all of the world had also made different responses. At that time, they just received one cloned sheep, and its rate was still low, some scholars in other areas had raised some doubts on it. However, as a major scientific breakthrough in the field, what' more, as a great applied prospect in biotechnology, not only British but also other countries' scientists continue to make efforts to tackle key problems, especially developed countries, and had obtained government funding. Not unexpectedly, since the cloned sheep "Dolly" was publicly reported a year, France and American scientists had reported they obtained cloned cattle. American scientists have obtained three cloned cattle with same genetic characteristics by nuclear transfer cloning biotechnology, the somatic cell they used was the fetal bovine fibroblasts, and then was transferred exogenous reporter gene by culture. Therefore, the nuclear transfer cloning biotechnology not only has been fully confirmed and had showed a great potential for the cloning of transgenic animal products.

3. Low-fat, high protein varieties of fishery animals will face a bright future

Because fish and other fishery products are a class of low-fat, high-quality protein foods with high-quality, they have entered into the international food safety system. Once in Rome, FAO held "World Fisheries Council of Ministers", and at the same year, in Kyoto, Japan, it held "the international conference: fisheries continue to contribute to food security". Both of the meetings highlighted the relationship between fisheries and food security, said the development of fisheries play a very important role in the food security of the world. Indeed, the world's fishery products have exceeded the production of livestock and poultry, and play the most important role in protein source for human. Therefore, most countries, including some of the states whose major fishing access depend on the natural fishery in the past time, begin to put emphasis on fishery research, and conducted with fishery biotechnology-related research. Particularly in the 21st century, sustainable development of fisheries has become a common concern of the world theme; scholars in the developed countries have reached an agreement about the role of aquaculture in the world fishery supply. Because the world is facing the threat of over-fishing[3], the sustainable development of fisheries has been the key to alleviate this crisis which is believed by international authoritative scholars[4-7], consequently, in developed countries, aquaculture has became the new rapid developing industry. The rapid development of aquaculture in China has changed the composition of the aquaculture, such as farming and fishing country in 1978, the ratio was 28:72, by 2007, it was 69:31, Chinese aquaculture production accounted for 70% of the world, in the field of aquaculture, China has become a truly great power.

Mariculture in China has undergone four tidal waves, such as algae, shellfish, shrimp, fish farming wave, the production of large seaweed, scallops, oysters rank first in the world, and shrimp and marine fish farming are also in the ascendant. However, compared with the livestock and poultry species, Chinese aquaculture, especially mariculture, the biological parents for breeding are most of the natural wild-type; their economic traits such as growth rate, disease resistance ability and output have great potential for genetic improvement. Therefore, the development of aquaculture genetic improvement biotechnology, the accelerating achievement of marine optimal varieties, especially in Mariculture animals (fish, shrimp, shellfish, its total production output of about 90%), the breeding of varieties with high quality and high resilience, will have a bright future[8].

4. Molecular design breeding will be the future of modern animal breeding biotechnology of livestock poultry and fishery animals

Molecular design breeding is system engineering depending on system biology, biological information and the genetics, the concept was produced just few years ago, but it is the future of genetic breeding research and breed improvement biotechnology. Compared with other important crops such as rice and wheat, the molecular breeding design of livestock, poultry and fishery animals is still in its infancy. The good news is that with complement of whole genome sequencing program of pigs, cattle, sheep, chicken and model fish such as zebra fish, medaka, puffer fish, and three thorns and the deepening research of genome function, especially a crop of characterized molecular markers and functional genes which are relating to the economic traits, there have had a large amount of basic data for the understanding of the regulatory network and its Economic Characteristics mechanism, meanwhile, in the field of functional genome and molecular design breeding, it provides important gene recourses for the further development of livestock, poultry and fishery animals. In conclusion, the fields of functional genome focusing on important economic traits and molecular design breeding biotechnology of livestock, poultry and fishery animals, have been called for more attention in developed countries, many countries have invested significant manpower and resources to carry out research, and it develops fast. China have done significant initial investment, and have a relatively solid ground, as a big livestock, poultry and fishery animals breeding country, China should be based on the past work, strengthen our animal aquaculture industry international leading position, and promote sustainable and healthy aquacultural development [9].

4.2 Biotechnology Development Goals

4.2.1 Development Features

Through analysis of the national needs and development trends of livestock, poultry and fishery animals, it can be summed up five main characteristics with the scientific and technological development of animal germplasm recourses and modern breeding:

1) The molecular and cellular biotechnology of genetic diversity and animal germplasm recourses evaluation, excavation, conservation and use, and assorted biotechnology system accordingly is the basis for the development of livestock, poultry and fishery products.

2) To livestock, poultry and fishery products, molecular design breeding is a key biotechnology for the improvement of output and quality.

3) Intensive farming is development direction of improving the output and quality of livestock, poultry and fishery products, and will exacerbate the pressure on animal disease prevention, therefore, healthy breeding and biosecurity biotechnology will be effective ways of promoting the sustainable development of livestock, poultry and fishery products.

4) To livestock, poultry and fishery animals, especially some deep-sea animals, the discovery of important functional genes and drug development will enhance the output of livestock and fishery animals and protect human health.

5) The ecosystem management of livestock, poultry and fishery animals, especially freshwater and marine germplasm resources, is sustainable development direction of livestock, poultry and fishery products.

4.2.2 Development Goals

Based on the above analysis, the next main objectives of animal germplasm resources and modern breeding biotechnology are:

1) To discover particular animal germplasm recourses of our country's important livestock, poultry and fishery animals, to establish important livestock, poultry and fishery animals germplasm resource sharing platform, protect important livestock, poultry and fishery animals germplasm resources, and strengthen ecological management; to ensure livestock, poultry and fishery products have a long-term sustainable and healthy development, and to enhance genetic improvement strength and potential innovation of livestock, poultry and fishery animals' fine varieties.

2) To explore functional genes and large-scale cloning with our own intellectual property rights, to establish the molecular design breeding biotechnology of important livestock, poultry and fishery animals, to breed new varieties of pig, cattle, sheep, chicken ,fish, shrimp, shellfish and other key breeding animals with quick growth, high protein content, high meatyield, high feed transformation or resistance to diseases; together with traditional breeding methods, to establish sex control breeding, disease resistance breeding, molecular design of intelligent animal breeding, molecular design of multi-trait breeding techniques for new varieties.

3) Based on a large number of the past obtained genetic resources with related to animal reproduction, breeding, growth and resistance, to carry out basic research of our important livestock, poultry and fishery animals functional genomics and molecular design breeding, through to carry out main economic characteristics of functional genomics research, such as reproduction, growth and resistance, etc., to clarify their gene regulatory networks and mechanisms theoretically, and to propose molecular design strategy of good breed; in technical methods, to establish Molecular Design breeding of multiple genes aggregation and gene manipulation techniques in important livestock, poultry and fishery animals, create high-quality breeding materials and the feasible molecular

design breeding approach of important livestock, poultry and fishery animals.

4) By analyzing the major pathogens through pathogen biology, pathogenesis, and epidemic law, to make effective ecology strategies to control fishery animal disease, and to develop suitable vaccines and effective drugs.

5) To unearth a crop of important functional genes of livestock, poultry and fishery animals and develop effective drugs with high value output and no harm to human health.

6) To achieve the ecological management of aquaculture; to promote health fisheries aquaculture, and protect water resource; to make marine and freshwater aquaculture with scientific layout , rational structure, abundant variety, complete specifications, high quality and energy saving.

4.3 Roadmap of the Scientific and Technological Advance

4.3.1 Key Scientific Issues and Tasks

In the future, the major scientific issues and tasks of animal germplasm recourses and modern breeding biotechnology of livestock, poultry and fishery animals include five areas as follow.

1. The molecular and cellular biotechnology of animal germplasm recourses evaluation, excavation, conservation and use, and their technical systems

Animal germplasm recourses of livestock, poultry and fishery animals are results of long-term development and evolution of life on earth, and are the material basis of human life, China is one of the countries rich in livestock, poultry and fishery animals germplasm resources, but the background and status of resources is still unclear, a number of important strategic biological resources not fully explore and play its role. The establishment of molecular and cellular biotechnology of animal germplasm recourses evaluation, excavation, conservation and use, and their technical systems will depend on the theoretical and technical development of functional genomics, systems biology and genetics. The key scientific issues to be solved include:

1) How does genetic basis of livestock, poultry and fishery animals germplasm recourses come to be?

2) How do molecular systems which affect livestock, poultry and fishery animal germplasm resources' specificity evolve?

3) How do the various environmental factors affect animal germplasm resources of livestock, poultry and fishery animals?

4) What is molecular and genetic basis of livestock, poultry and fishery animals' adaptive evolution?

5) Animal fishery animals to interact with the environment, the role of the

mechanism?

In response to these technological issues, the main future research tasks of this area are:

1) Explore of strategic livestock, poultry and fishery animals' germplasm recourses: to invest and explore special animal germplasm resources of extreme environments, especially in the deep sea not yet known for human; to develop different molecular markers, and comprehensive evaluate economic use and development prospects of important strategic animal germplasm resources.

2) Development of identification of different livestock, poultry and fishery animals' DNA barcode: Through the establishment of germplasm recourses' evaluation, preservation and use of shared platforms, to fully explore and use our abundant resources of livestock, poultry and fishery animals; Using genome sequencing and bioinformatics research to distinct and identify different livestock, poultry and fishery animals' DNA barcode; to develop practical DNA barcode biotechnology to quickly identify the pig, cattle, sheep, chickens and other livestock or poultry as well as different varieties of fish, shrimp, shellfish and other major aquaculture species.

3) Establishment of somatic cell cloning and single-sex reproduction biotechnology of livestock, poultry and fishery animals: to develop biotechnologies of livestock, poultry and fishery animals germplasm recourses' evaluation, excavation, conservation and use, such as molecular marker techniques, adult stem cell cloning, parthenogenesis and germplasm integration biotechnology, etc., and enhance genetic improvement strength and potential innovation of livestock, poultry and fishery animals.

4) The study of important livestock, poultry and fishery animals' genetic basis of phenotypic diversity and the formation mechanism: to investigate the genetic basis of phenotypic diversity and the formation mechanism of important livestock, poultry and fishery animals, and provide the appropriate scientific basis for the effective conservation and utilization measures of livestock, poultry and fishery animals.

2. Livestock, poultry and fishery animals' molecular design breeding biotechnology and the related technical systems

Molecular design breeding of livestock, poultry and fishery animals and the related technical system rely on the establishment of systems biology, functional genomics, bioinformatics, genetics, breeding ecology and other aspects of knowledge, the main scientific and technological issues includes as follow:

1) How to use the reproductive strategies and characteristics of livestock, poultry and fishery animals to cultivate the fine varieties with strong reproduction or gender controlled characteristics by molecular design breeding biotechnology? The key is to understand the gene regulatory networks and their interactions of livestock, poultry and fishery animals' reproduction.

2) How to through the growth functional genomics to cultivate the fine varieties with quick growth characteristic by molecular design breeding biotechnology? The key point is to reveal the control master genes and gene regulatory

networks, as well as the coordination mechanism between reproductive and growth, which can regulate the growth of livestock, poultry and fishery animals.

3) How to effectively design and cultivate new livestock, poultry and fishery varieties with a specific resistance characteristic? The fundamental is to grasp the disease resistance mechanism of immune of livestock, poultry and fishery animals.

4) How to cultivate new livestock, poultry and fishery varieties with high feed transformation characteristic? The key is to understand livestock, poultry and fishery animals' nutritional needs and the gene regulatory network and mechanism of its digestion and absorption.

5) How to achieve molecular design breeding objectives and technical approach of livestock, poultry and fishery animals? First is to achieve the multi-gene aggregation biotechnology of quantitative trait; second is to design the safe genetic regulatory elements, and establish a stable and efficient gene operation biotechnology of livestock, poultry and fishery animals.

In response to these technological issues, the main future research tasks of this area are:

1) Study of functional genomics and gene regulatory networks of important livestock, poultry and fishery animals' main economic traits: to study the functional genomics for livestock, poultry and fishery animals' reproduction, growth, milk, immunization, disease or stress resistance research, and reveal the gene regulatory networks and its mechanism, as well as the interactive regulatory relationship between different economic trait networks, to clone and identify functional genes with important breeding value, to provide theoretical foundation and genetic resources for the innovation of molecular design breeding.

2) The technical basis research of important livestock, poultry and fishery animals' molecular design breeding: to develop sex control molecular design breeding, multi-trait molecular design breeding, intelligent molecular design breeding and multifunction molecular design breeding biotechnology; together with traditional breeding methods, to breed new varieties of pig, cattle, sheep, chicken ,fish, shrimp and shellfish with quick growth, high protein content, high meatyield, high feed transformation or resistance to diseases.

3) The research of pathogenic mechanism of important livestock, poultry and fishery animals and the cultivation of new livestock, poultry and fishery varieties with disease resistance: through the identification of pathogen, pathogenesis, interaction between pathogen and host and its interaction mechanism, to develop specific pathogen' efficient vaccine producing biotechnology, efficient drug developing and prevention and control biotechnology, non-pathogenic breeding of immune control biotechnology, eco-culture environment regulating and eco-efficient management biotechnology to achieve efficient, non-intensive farming and to provide health and safety of livestock, poultry and fishery animal products.

3. Animal disease prevention and control strategies and biotechnology of livestock, poultry and fishery animals

The precondition of livestock, poultry and fishery animal disease prevention and control is to understand gene regulatory networks and its action mechanism of the animal's immune response, and then on this basis, analyze the application value of immune regulation on animal disease-resistance to the fine breeding livestock, poultry and fishery varieties. The key biotechnology is most worthy to study includes:

1) What are livestock, poultry and fishery animals' specific and non-specific immune system gene regulatory networks and their interaction mechanisms?

2) What are the important livestock, poultry and fishery animals' pathogens and host immune system and its interaction mechanism?

In response to these technological issues, the main future research tasks of this area are:

1) The functional analysis of gene regulatory networks of livestock, poultry and fishery animals' specific and non-specific immune system and their interaction mechanism research: to analyze the gene regulatory networks of livestock, poultry and fishery animals' specific and non-specific immune system, and illustrate the function and mechanism; to reveal the relationship of the two systems and the channels, signal pathway and mechanism of the interaction.

2) The research of important livestock, poultry and fishery animals' pathogens and host immune system and its interaction mechanism: aiming at the important animal pathogens of livestock, poultry and fishery animals, such as mad cow disease, swine foot and mouth disease, avian flu, fish hemorrhage, fish lymphatic cyst, etc., to carry out studies of the pathogen and host immune system' interaction and its mechanism, to clarify the signal regulation network and the disease-resistance mechanism.

4. Livestock, poultry and fishery animals' important functional genes exploring and drug developing Biotechnologies

The conservation and use of Livestock, poultry and fishery animals' genetic resource has become the focal point of worldwide attention. In addition to traditional use of animal resources, because of the rapid advances in biotechnology, people begin to explore unknown biological resources in extreme environments and constantly find new species, especially animals live in extreme environments such as deep-sea, and their genetic resources have started to become scientists and Investors' concerned target. The potential scientific and technological issues and research tasks include:

1) Researches of exploring Important functional gene and developing drug biotechnology: to establish efficient gene transfer system and the expression system for special animal genes, using efficient expression system such as engineered bacteria and biological reactors to produce functional gene products as feed, drugs, diagnostic reagents etc; to achieve expression and application of

breeding animal' functional genes with important traits.

2) Special animal genetic resources discovery and drug development in extreme environments such as deep-sea: to establish selection and utilization biotechnology of special animal genetic resources in extreme environments such as deep-sea; to develop and utilize new marine drugs and health care products for major diseases such as renal failure, cardiovascular, anti-bacterial, anti-aging; to develop new and efficient fish gene vaccine, by using new Biotechnologies of gene recombination and large-scale production, to develop new environmental and friendly antibiotic alternatives - antimicrobial peptides; to develop new marine medical materials (such as nano-materials), and new marine biological pesticides (including insecticides, fungicides and acaricides); to develop new plant growth regulator.

5. The ecological theory and biotechnology of fishery animals' germplasm resources protection

With the worldwide decline of biological diversity, habitat loss, and increased demand for precious biological resources, biological diversity, particularly featured species protection of rivers, lakes and marine areas, has become the focal point of worldwide attention.

In the field of the ecological theory and biotechnology of fishery animal resources protection, the key scientific issues and researches mainly in the following areas:

1) Large survey of freshwater and marine fishery animal germplasm resources: Through the establishment of featured species list and protected areas, using electronic scanning and tracking biotechnology, biological cloning and molecular ecology biotechnology and molecular phylogenetic analysis biotechnology to do large survey of freshwater and marine fishery animal germplasm resources, to establish standards of the specimens, samples (DNA, tissue , blood samples, frozen samples, etc.) and collected data, to achieve the standardization, maximize and digitalization of collected information.

2) The establishment fishery animal germplasm resources database and dynamic monitoring system: to carry out a comprehensive and in-depth scientific investigation in key areas and important stocks, and learn more about animal resources' background and current situation within the region; to reveal the rich and complex biodiversity of the area as well as the region's role and special status in the country and the world's biological diversity; to establish fishery animal germplasm resources database and dynamic monitoring system.

3) The establishment of national important fishery animal germplasm resources library and popular science exhibition hall: through the use of modern aquarium construction and management techniques, establish important fishery animal germplasm resources library and science exhibition hall of freshwater and seawater, collect different species and different varieties germplasm resources of important fishery animals, and thus to carry out breeding, genetic operations and improvement, make it to be a platform and base for research and

popular science.

4.3.2 Development Roadmap and Design Ideas

In view of still relatively low supply of livestock, poultry and fishery animal protein and there would be still two to three times needs in future, the situation is very grim, focus on the future, we mainly use integration of multidisciplinary research methods in the fields of life science and biotechnology such as systems biology, integrative biology, bioinformatics, genomics and proteomics, genetic engineering, etc, with major product-oriented research & development strategy, develop healthy and sustainable development of animal aquaculture; Through the establishment of germplasm recourses' evaluation, preservation and use of shared platforms, develop biotechnologies of livestock, poultry and fishery animals germplasm recourses' evaluation, excavation, conservation and use, such as molecular marker techniques, adult stem cell cloning, parthenogenesis and germplasm integration biotechnology, etc., and enhance genetic improvement strength and potential innovation of livestock, poultry and fishery animals.

Though revealing the molecular mechanism of livestock, poultry and fishery animals' reproduction, growth, milk, immunization, disease or stress resistance, we will clone and identify functional genes with important breeding value, and develop sex control molecular design breeding, multi-traits molecular design breeding, intelligent molecular design breeding and multifunction molecular design breeding biotechnology; together with traditional breeding methods, we will breed new varieties of pig, cattle, sheep, chicken, fish, shrimp and shellfish with quick growth, high protein content, high meatyield, high feed transformation or resistance to diseases.

Through the identification of pathogen, pathogenesis, interaction between pathogen and host and its interaction mechanism, we will develop specific pathogen' efficient vaccine producing biotechnology, efficient drug developing and prevention and control biotechnology, non-pathogenic breeding of immune control biotechnology, eco-culture environment regulating and eco-efficient management biotechnology to achieve efficient, non-intensive farming and provide health and safety of livestock, poultry and fishery animal products.

On the basis of identifying functional genes with drug value of livestock, poultry and fishery animals especially the deep sea animals, we will establish functional gene discovery and drug selecting platform, carry out research & development of efficient, specific featured functional drug and biological products, and enhance livestock, poultry and fishery products' use value, and then achieve scale industrialization and high value of economy.

Through revealing the ecological basis of important fishery animals' propagation preservation, we will use fisheries structural adjustment with relation to ecological principles and biotechnology, develop healthy and sustainable aquaculture model, and achieve the healthy and harmonious sustainable development of freshwater aquaculture and mariculture (Fig. 4.1).

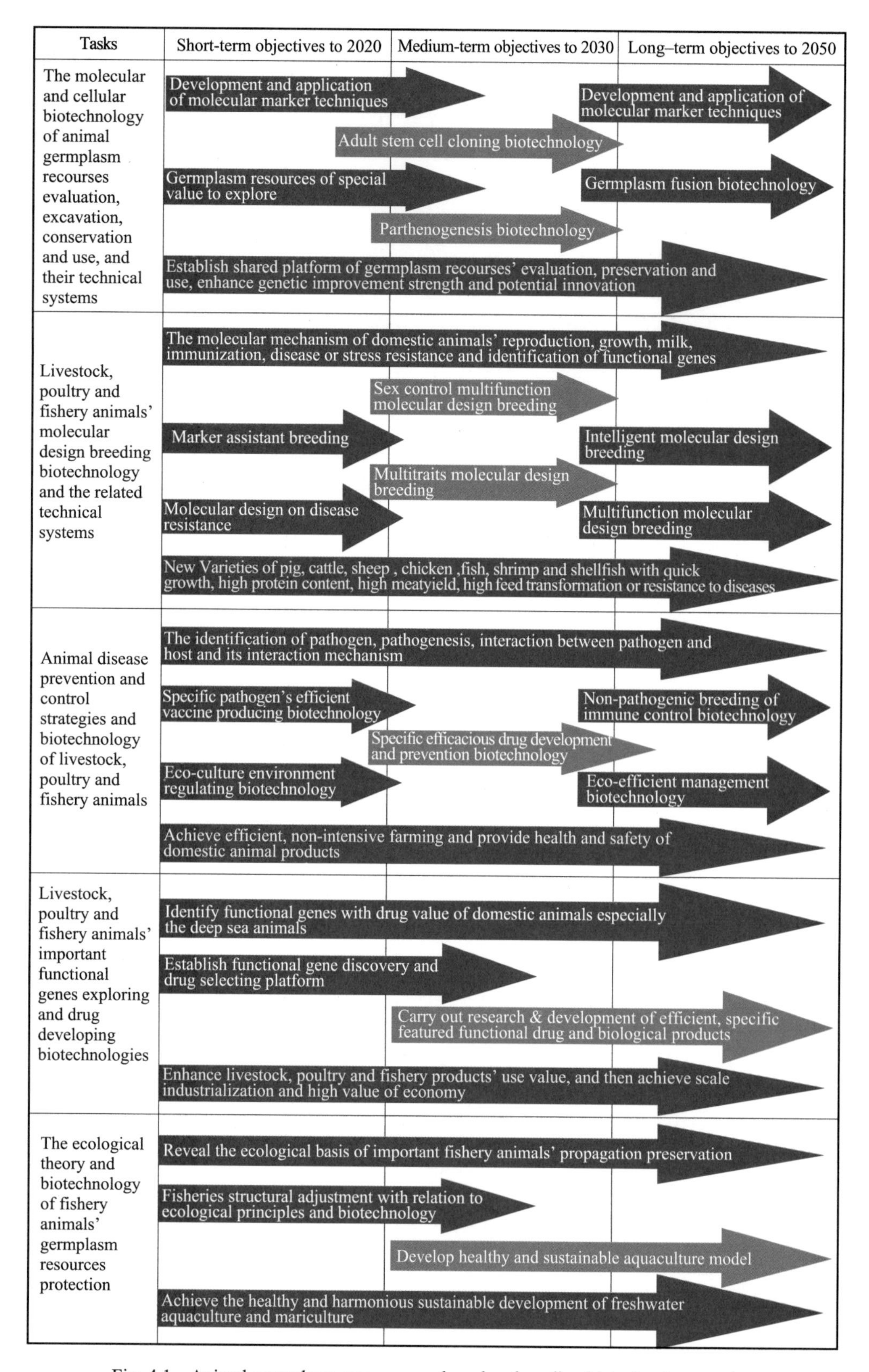

Fig. 4.1 Animal germplasm resources and modern breeding biotechnology roadmap

Main References

[1] Duan CC. Who will feed China is still a question. Global Times, 2008.

[2] Wu CX. Animal agriculture genetic breeding problems in the twenty-first century. China Science and biotechnology Information. 2004, 17: 6-8.

[3] Worm B, Barbier EB, Beaumont N, et al. Impacts of biodiversity loss on ocean ecosystem services. Science, 2006, 314:787-790.

[4] Naylor RL, Goldburg RJ, Primavera JH, et al. Effect of aquaculture on world fish supplies. Nature, 2000, 405: 1017-1024.

[5] James HT, Geoff LA. Fishes as food: aquaculture's contribution. EMBO reports, 2001, 21: 958-963.

[6] Pauly D, Christensen V, Guénette S, et al. Towards sustainability in world fisheries. Nature 2002, 418: 689-695.

[7] Pauly D. The Sea around Us Project: documenting and communicating global fisheries impacts on marine ecosystems. Ambio, 2007, 36: 290-295.

[8] Wang QY. The basis research and focus areas of mariculture biological resources. Science Foundation of China, 2005, 6: 334-338.

[9] Gui JF. The present and future of genetic and basic development research of fish Variety improvement. Life Sciences, 2005, 17: 112-118.

5 Roadmap of Resource Saving Agricultural Science and Technology Development

Agriculture development in our country faces the limited arable land resource which is further continuously reduced. The increasing shortage of land resource restricts the agriculture development. This difficulty is, moreover, intensified by acute shortage of water resources and fierce competition between agricultural departments and non-agricultural departments. Meanwhile, in our country intensified agriculture ecosystem with high input and high yield will remain the most important ecological mode for agricultural production in the future. Therefore, future agriculture development not only needs fundamental breakthroughs on increasing production technology, but also needs the upgrading technology on resource saving agriculture and arable land production capacity. These technologies may support and secure the transformation of agriculture from traditional model to resource saving model.

5.1 Development Requirement, Significance and Tendency

5.1.1 Requirement and Significance of Technological Development

1. Faced with severe shortage as well as degrading quality of arable land resources, the government should devote energetic efforts to alternative technologies for increasing arable land area, improving production capacity, and establishing a land-saving agricultural production system

Scarcity of arable land resources and the degrading quality have already become the major restraining factors for agricultural production. By the end of 2008 the total arable land area of the whole country was 1.826 billion mu with a per capita coverage of 1.38 mu (total population was 1.328 billion by the end of 2008). The net reduction amount was 290,000 mu, which was 50% of that in

2007, indicating that the tendency of arable land reduction has been basically curbed. In recent years, although the government has adopted measures in protecting the arable land from loss through judicial land management and new arable land resources exploitation, newly reclaimed land in place of occupied fertile land has resulted in a fall in production capacity. In addition, the national agriculture development is also faced with increasingly soil degradation problem[1]: ① Accelerated arable land degradation. High and stable yield land decreases, and medium- and low- yield land accounts for 2/3 of the total. ② Serious soil erosion. Soil erosion area of cropland has reached approximately 48.67 million hectares, making up 38 % of the total arable land area. ③ Large contaminated arable land. According to incomplete statistics, approximately 150 million mu arable land in our country has been contaminated to varying degrees, in which 32.50 million mu has been contaminated by sewage irrigation, 2 million mu by solid waste and physical damage. They mostly happen in relatively developed area.

2. With unbalanced supply and demand on water resources and uneven spatial and temporal distribution of water resources in our country, the situation of water resources is severe in future, and there is an urgent need to establish a water-saving and high-efficient agricultural production system

The agricultural production in our country has been threatened by the imbalance between supply and demand, uneven spatial and temporal distribution, as well as low utility ratio of water resources[2-4]. With total amount of water resources reaching 2,800 billion m^3 in our country, per capita share of water resources was only 1,927 m^3 in 2006, which is only 42% that of 1950, close to the international alarm level of 1,700 m^3. Affected by monsoon climate, rainfall mostly takes place in summer, which easily increases the odds of draught in spring but flood in summer. The dramatic variation in rainfall often results in serious floods and low water of rivers. At present, with uneven distribution of water and soil resources, the Yellow River, the Huaihe River, and the Haihe River basins are most known for their water stress. In the three river basins, the land area accounts for 13.4% of the total in China, arable land is 39%, population is 35%, but water resources is only 7.7%, with less than 400m^3 per mu for arable land. However, inland river basins in northwest China have 35% of total land area, 5.6% of total arable land, 2.1% of total population and 4.8% of total water resources in our country. Northern China owns more than one half of wheat output and 1/3 rice output of our whole country, and its total grain output accounts for more than 1/3 of national output, but the water resources per capita accounts for only 1/7 of the country. Yangtze River in south China and Northeast China are the main production bases for rice. Under global climatic change, a number of river basins such as Yangtze River and Yellow River are now faced with the following problems due to the decreasing supply from melting glacier of riverhead: drastic runoff decrease in low water period, increasing flood disasters, and increasingly difficult irrigation. The risks and threats posed by abovementioned problems all lead to crop yield reduction[5].

On the other hand, the utility ratio of water resources is low. The utility ratio of agricultural irrigation water in our country is only 40-50%, while in developed countries it can achieve up to 70-80%. Grain output per m^3 water is about 1/3 of that of international level. Therefore, our country should exert great efforts immediately to promote water-saving irrigation and high-efficient utility of water resource technological system and establish water-saving and highly efficient modern agriculture.

Soil Degradation

As the core of land degradation, soil degradation is referred to as a process in which soil quality and its sustainability decreases even totally loses due to many natural and human factors. Soil quality includes three aspects: to maintain production capacity of the ecosystem, to keep land use and environmental management sustainable, and to promote animals and plants healthful.

Soil degradation in China can be classified into 7 types: namely, soil erosion, soil desertification, soil salinization, soil impoverishment, soil gleization, soil contamination and soil fertility loss[6,7]. Total soil erosion area in our country has risen from 1.5×10^8 hectares in 1950s to 1.7×10^8 hectares nowadays. Desertified land area in our country is 3.34×10^7 hectares which is increased with 2.1×10^5 hectares per year. Land area subject to salinization-alkalization is 3.69×10^7 hectares. The second national general survey of soils indicated: the average of soil organic matter was less than 1.5%, the phosphorus-deficient soil area in the country rose from 2.7×10^7 hectares in 1953 to 6.7×10^7 hectares in 1995, gley soil area was approximately 2.08×10^7 hectares, and contaminated soil area increased from 1.65×10^6 hectares in 1983 to 8.65×10^6 hectares in 1995 (accordingly rising from 1.66 % to 6.26 % of total arable land).

3. High-intense agricultural ecosystem with high input of resources and energies is still a principal ecological mode for sustainable agriculture in the future. The government must promote efficient use of nutrients, water and energy resources, and establish fertilizer-saving and energy-saving agriculture ecosystem

Nowadays, in order to meet the increasing demands of grains due to population growth and improving living standards, the intensified agriculture ecosystem with high inputs and yields still remains the principal ecological mode for agricultural production. But it has already triggered many serious environmental problems, resulting in a bottleneck for agricultural development in our country. Occupying 8.9% of the world's total cultivated land, China succeeds in producing 21.1% of the world's total grain. However, this success is at the cost of consuming 35% of world's total fertilizer and large amount of pesticides, water and energy resources. Presently, the amount of fertilizers and pesticides applied in our country ranks the largest in the world, with the annual consumption of nitrogen synthetic fertilizer (pure nitrogen) and pesticides exceeding 25 million ton and 1.3 million ton, respectively. The amount applied in a unit of area is 3 times for nitrogen and twice for pesticide than those of world's average. In 2006, the total amount of water used for agriculture in our country reached 366.44 billion m^3, accounting for 63.2% of total water used in the whole country. Auxiliary energy used in machinery during the agricultural development increased, for example, the annual power for tractors tillage and the power for other agricultural purposes exceeded 4,900 million kW and 2.2 trillion kW. However, with low utilization rate of agricultural resources, the fertilizer effect (the amount of rice yield by a unit of fertilizer) decreased year by year, with 22.9 kg/kg in 1990 down to 15.9 kg/kg in 2003. The utilization rate of pesticides on average is possibly only around 30% in our country. Large amount of unused fertilizers and pesticides were left in the farm field, or transported into surface water and ground water, resulting in a series of environmental problems[8]. Water resources and water quality has been deteriorated to a grave extent due to over exploitation, with utilization rate of water resources exceeding 80% in the Yellow River, the Haihe River and the Huaihe River[9]. The lower reach of Yellow River has been subject to no flow every year since 1990s. Meanwhile, ground water in northern part of China is over pumped for agriculture and the ground water depression cone has become rather serious. In the Haihe ground water table has been drastically decreased with over exploitation up to 60 billion m^3. In the North China Plain the area under over exploiting ground water increases from 87,000 km^2 at the beginning of 1980s to 190,000 km^2. The deep ground water table in Hebei Province decreases by 2 m annually on average.

Until the middle of 21st century, the population in our country will keep increasing. Then, the crop yield of per unit will be requested to increase by over 20% even if the current grain consumption rate per capita is maintained. In addition, the increasing demand for other agricultural productions will lead to an increase of agricultural chemicals consumption, which will definitely exert more pressure on agricultural environment. Hence, the government should adopt a sustainable development method which integrates quality, efficiency, sustainable yield and environmental protection, and find a solution to high-efficient utilization of water, nutrients and agricultural chemicals. Enhancement of resource utilization rate by way of researches on related technologies will be a long-term goal on which the national agricultural development depends. In order to realize this goal, the substitution strategies for scarce resources and the high-tech strategies for resources management must be implemented besides overall upgrade of existing technologies. Some alternative resources should also be utilized, for example, exploiting grassland, water body and sea for more carbohydrate and protein in place of arable land, so as to expand tangible and intangible scarce resources indirectly. Agricultural resources should be monitored, forecasted, optimized and operated precisely through agricultural information technologies with an aim to enhancing the management benefits of resources.

5.1.2 Tendency for Technological Development

Resource saving agricultural development includes two meanings: the first is saving, and the second is intensifying. Resource saving agriculture is actually a type of intensified agriculture with intensive cultivation, high-efficient output, high income, and sustainable development. In the recent 50 years, the priorities of agriculture in the world have been to develop land saving agriculture, water saving agriculture and energy saving agriculture in terms of resources saving agriculture. It has a multi-croppings and multi-levels, as well as has the characteristics of high-yield and low-input intensified agriculture, including adopting advanced irrigation system, irrigation technologies and scientific fertilization. As the natural environment and resource conditions vary from one place to another in our country, the production system of resource saving agriculture should be developed based on the regional resource status and the needs of sustainability so as to transform the traditional agriculture to modern agriculture.

1. Establish policy and measure system for intensified utilization of arable land resources and fertilization technology system for arable land quality improvement

Since 1960s, based on the development of information technology and automatic monitoring technology such as satellite remote sensing, developed counties, including Europe and America, have developed the monitoring and forecasting technologies, and have established the network monitoring the long-term change of arable land at national scale. For instance, the network of ecosystem at national scale and the platform for water and soil quality of arable land at regional scale were established in America in 1980s and in Europe and Canada in 1990s. Recently, intelligent wireless network monitoring system and distributed data collection and management platform have received much attention.

Since the beginning of 1950s, our country has begun to devote itself to the researches of regional arable land planning and of the assessment of comprehensive bearing capacity of arable land. In 1980s our country started to the research on the intensified technology for arable land at regional scale. In 1990s our county began to pay more attention to land saving technology and planning technologies applied in arable land. Presently, to establish the "arable land substitution technology" has become a trend at different regions, including land replenishment, land exploitation as well as land reclamation. It is also to start large-scale management of regional arable land, and to establish overall safeguard and regulation system of land resource at national scale[10].

General soil surveys were conducted twice in 1950s and 1980s, at national scale, respectively. Long-term researches on soil fertility evaluation and evolution were carried out at various agricultural zones. Some technological systems (e.g., biological, chemical and physical, etc) were established for restoring and rehabilitating different types of degraded soils (erosion, acidification, nutrients depletion, non-equalization, biological function degeneration, desertification, salinization and alkalization), and different types of impedient soils (including physical, chemical and biological obstacles). Since 1990s, soil fertility improvement technology has been transformed into the nourishing technological system for improving soil environmental quality and health quality. Our country began to do researches on soil quality evolution and oriented nourishing technology for red soil, black soil, Chao soil and paddy soil[1]. In the near future, methods for evaluating and monitoring arable land quality and technologies for enhancing soil quality and fertility steadily need to develop.

Resource Saving Agriculture

Resource saving agriculture can be classified into three modes, including land-, water- and energy-saving agriculture[11]. Land-saving agriculture refers to a type of system with suitable crop rotations and/or multiple vertical crop levels so as to increase the absorption of light and thermal resources and utilization of physical energy. The crop systems are constructed in light of the principle of symbiosis and complementation among species and growth attributes of various species and based on temporal and spatial difference during their biological growth. Hence, the aim of land-saving agriculture is to enhance the unit yield rate, including various forms and levels of stereo agriculture, e.g., intercropping, multiple cropping. If the multi-crop index is increased by 1%, the increase of yield will be equal to that of 1.3 million hectares of arable land.

Water saving agriculture is a kind of system in which it makes full use of rainfall and water resources available, adopts a number of efficient water-saving measures in order to enhance water use efficiency. It is composed of water-saving irrigation agriculture and upland water-saving agriculture (rain-fed agriculture).

Energy saving agriculture can be illustrated at two aspects: firstly, to make the best use of conventional energy resources, such as coal, oil and natural gas, and secondly to enhance the utilization rate of biological resources. For example, ecological agriculture could be defined in light of various patterns, including biological cycling, comprehensive use of waste, and symbiosis among different species as well as three-dimensional agriculture which makes the best use of space.

2. Develop water saving technological system for agricultural ecosystem and establish water resources safeguard system

In the recent 50 years, technological measures for water-saving irrigation can be summarized as follows: improve irrigation support system and management; enhance canal-system water use efficiency; adopt advanced irrigation technologies and reduce water loss; establish water-saving irrigation mode; and implement water-saving irrigation regulation. Water-saving irrigation system in the future should integrate water transport system with minimum water loss, automatic control system (waterhead allocation, soil moisture status forecasting and field irrigation), and technology and measure system. The comprehensive technological system for the upland water-saving agricultural development includes many measures as followed: adjust agricultural structures and cultivar in light of water and soil; to choose, cultivate and extend anti-drought and high-yield species; to reform farming system; to enhance soil moisture saving ability; to improve soil fertility by applying more organic fertilizer; to enhance the utilization rate of rainfall; to make use of close agriculture climate engineering to prevent soil evaporation. In terms of water saving chemical technology, to develop anti-evaporation chemical agent prevents soil evaporation, reduces crop transpiration, and enhances water producing ability of soils. Recently, in the field of anti-drought and high-yield species, the common international practice is to take advantage of genetically modified crops to enhance water utilization rate and promote "Blue Revolution"[12,13].

Recently, water saving technology developed in our country includes researches on high-efficient water regulating system when crops are harmed and then self-repaired upon limited water deficit; researches on water saving in different varieties, chemical water saving, limited irrigation principles and related technologies based on conventional agricultural technologies such as improving soil fertility, anti-drought to maintain soil moisture, as well as film and straw covering; researches on management technology of water and fertilizer coupling principle. It is to do researches on the irrigation mode combing channel and well in connection with channel fed area, and promote spray irrigation and drop irrigation and their supporting facilities based on surface water saving irrigation technological system (like flat ground irrigation, furrow irrigation and intermittent irrigation), establish forecasting system and management platform for water resources use, as well as integrate water saving technological system (engineering, biology and chemistry) at river basin scale.

In the world, technologies and policies on water balance and water distribution have received much attention. In reaction to occurrence of water crisis like water shortage and water pollution, many counties started to bring forward the corresponding management measures for water resources in 1960s. They should put forward the development plan for water resources at national scale, safeguard the freshwater resources, do researches on information-based, data-based and model-based management pattern and measures for water resources, and achieve a rational allocation of water resources through certain regulation measures such as supply and demand relationship and pricing lever. At present, facing with water crisis for agricultural purpose, the key researches are food security and water quality. The demand and consumption rules of water resource, safeguard measures of water resources, new water-saving technology exploration based on evaporation transpiration (ET) management need to develop. The joint development and utilization technology for surface water and ground water, particularly agricultural water-saving society based on river basin knowledge management (KM) are also critical processes. Nowadays, our country has shifted its main investment in terms of agricultural water projects from establishing engineering projects and new irrigation system to establishing water-saving and high-efficient agriculture, intensifying technological platform for water resources security and the construction of technological support capacity, carrying out researches on water demand forecasting for agriculture in response to global climatic changes, as well as innovating comprehensive management of river basin water resources.

3. Develop new high-efficient fertilizer and integrate high-efficient utilization measures of nutrients and energy in agriculture ecosystem

In the recent 50-60 years, fertilizer development has moved towards various directions, including the composite and high-efficient, slow/controlled release and environmental friendly fertilizers. Referring to developed countries, the overall trend for fertilizer evolution is to develop high-efficient compound fertilizer, and to reduce ingredients with side effects, in order to meet the needs of high yield, high efficiency and high quality, to cut the cost of packaging, transportation, storage and application, and to enhance soil fertility[14,15]. Nowadays developed countries most rely on compound fertilizer. For example, in America and the United Kingdom, compound fertilizer accounts for 79%. In Japan, France and Germany, it is 60% to 80 %. In recent years, developed countries like USA have begun to produce super concentrated compound fertilizer, such as APP (16-62-0) and PTPP (0-57-37); to produce multiple nutrients compound fertilizer containing moderate-element such as calcium, magnesium and sulphur, and to prepare multi-functional compound fertilizer containing organic matters, growth hormone, herbicides, pesticides and trace elements.

The controlled-release fertilizers have received much efforts in the world in the past few years, including biochemical- inhibitor slow release fertilizer (like dihydric phenol, chinone and MAP compound), low water-solubility inorganic or organic compound fertilizer (like inorganic ammonium phosphate compound, urea formaldehyde and oxalamide), and coated controlled-release fertilizer. Although foreign controlled-release fertilizers have many manufacture, they remain relatively high price and low market share, and then consumption is low on the whole. There are generally several problems of controlled-release fertilizers: ① as for agricultural plants, it is found to be really difficult to meet the need of plants in their early stage or during peak period of fertility absorption; ② as the price of some fertilizers (like high polymer coated fertilizer) remains too high, it is difficult for them to satisfy the need of field crops, and they are more often used on horticulture crops (fruit plants) and other perennial plants known for their ornamental value; ③ some controlled-release fertilizers is still not quite meet the standards of environment, especially the requirement of sustainable development. In the recent years, based on the successful application of hydrophilic polymer materials as controlled-release carrier in the field of medicine and agricultural chemicals, American researchers have invented a new concept of fertilizer[16]: gel fertilizer, which uses hydrophilic polymer materials as a carrier for controlled-release of nutrients. The hydrophilic polymer materials represent a new direction for developing new controlled-release fertilizer, which have some advantages like degraded by soil microorganism, small amount of additive and low cost.

Our country has begun to develop ammonium phosphate compound fertilizer since 1960s, and presently, China is increasingly tending to adjust the proportions of N, P and K in light of balanced nutrients supply, and reasonably mix moderate and trace element based on regional climate-soil-crop conditions; secondly, it is to develop compound fertilizer, and work on suitable equipment and techniques for local farmers in light of regional soil test at the county scale, fertilizer formula as well as research on supporting system of regional farm nutrient resources management; it is also to develop new slow/controlled release fertilizer for cash crop, fruits and vegetables, and eventually set up resource-saving and high-efficient agriculture ecosystem[17].

To deal with problem of high consumption of fertilizer in the intensified agriculture, some integrated technologies have been proposed e.g., high-efficient fertilizing system for farm field, comprehensive management and supporting technologies for multi-nutrients, assorted mechanized fertilization technology during the overall production process, in light of regional soil nutrient situation and crop demand law under high-yield mode. Since 1980s, agricultural environment has received much attention in the world. As guided by the principle of energy and matter cycle in agriculture ecosystem, nutrient-recycled and multi-level biological utilization technologies has been established; it is also to exploit agricultural biology energy to much effort[18]; and to develop high-efficient utilization technology of biological resources[19], in which the key point is crop straw burning control and substitution technology, including fast microbiological decomposing technology for farm residues and residue burning substitution technology and equipment.

In terms of highly-efficient use of farm nutrients and energy, on one hand, high-yield farm crop and nourishing technology easily decrease the utilization rate of fertilizer and increase the risk of nutrient loss in the high-input mode. On the other hand, renewable resources such as straw, branch, leaf, and foliage from farm field, are not made full use of. Hence, an efficient utilization of nutrients from farm fertilizer and organic wastes has been received much attention, while all efforts have been made to promote minimal or zero tillage at cost of low energy and improving soil and water conservation. With the fast

Biomass Resources and Utilization Technology

Biomass can be defined as various forms of organisms directly or indirectly derived from photosynthesis, including all the animals, plants and microbes. With abundant resources in biomass, our country has around 5 billion ton biomass resources theoretically, 4 times of that of current national consumption. The main sources are as followed: agricultural wastes (such as straw, rice husk and sugar cane residues), forestry wastes (such as waste timber), industrial wastes (such as wastes from paper mill and food factory), household wastes, organic waste water (such as human and animal excreta, urban sewage, and industrial organic sewage water), energy crop (such as sugar cane, cassava, oil rape and sweet sorghum) and energy plant (such as fast-growing forest and Chinese silvergrass). In 2004, China has actual reserves of biomass resources as following: 728 million ton of straw, 3.926 billion ton of animal excreta, 2.175 billion ton of forestry biomass, 0.155 billion ton of urban wastes, 48.240 billion ton of sewage water. By 2020 the amount of biomass resources estimated will reach up to 2.93×10^{19} J[20].

In China, the recycle technologies of agricultural and forestry biomass resources in farm ecosystem has always been received much attention, e.g., straw direct returning, fast composting and returning (concerned with related microbiological technologies) as well as organic fertilizer manufacturing technologies. In terms of energy utilization, it is mainly to develop biomass cultivation, transforming technology and equipment, such as straw gasification instrument, fuel gas purification technology, heating system technology depending on direct burning of straw and timber, ethanol production technology based on cellulose, as well as modular methane fermenting unit and supporting technology.

growth of social economy in our country, agricultural cultivation mode moves towards labor saving and time saving mode. Meanwhile, the amount of chemical fertilizers input increases and lump-sum input becomes common. Therefore, the balance of supply technology for farm nutrients becomes important[21].

Since 1990s, the best management practices for water and fertilizer (BMP), and the coordination between output, economic benefits and environment, have become an important aspect of sustainable agriculture. Regional fertilization decision-making is a principal part of BMP research, focusing on nitrogen management. Recently, with the increasing demand of agricultural production and regional environment management, researchers have established regional decision-making fertilizing supporting system based on process model, GIS, database technology, such as SUNDIL 2.0 and AFOPro 2.1 from Rothamsted Experimental Station, United Kingdom. Model application is moving towards shared network development.

5.2 Technological Development Goal

5.2.1 Overall Goal

It is to build a perfect monitoring and forecasting platform of national arable land and water resources, a perfect management technology research platform of water and nutrients, and a perfect new fertilizer research and development platform. It is to establish all-round three agricultural production technology systems including land-saving agriculture, water-saving agriculture and fertilizer- and energy- saving agriculture. It is to enhance all-round new fertilizer industry and modern agricultural equipment industry. It is to realize intensified utilization and management of regional water and soil. It is to implement agricultural mechanization by way of water-saving irrigation and highly-efficient fertilization. It is to standardize precision water and fertilizer management and energy-saving minimal and zero tillage.

It is to secure that arable land area with medium and low yield should be reduced by 50% to 60%. The comprehensive utilization rate of soil, fertilizer and water in agriculture ecosystem should be improved by 30%. The nutrient and energy input should be reduced by 25% to 30%. It is to popularize intelligent fertilizer, to realize the dynamic balance between arable land and water resources on which food safety production depends. It is to set up modern agricultural production system with a high and stable yield, high efficiency and high quality in order to realize sustainable agriculture.

5.2.2 Goals of Different Stages

1. Short-term goal in 2020

It is to set up the monitor and forecast platform of arable land and water resources, comprehensive management and research platform of water, fertilizer and tillage as well as new high-efficient fertilizer research and development platform at regional scale.

It is to implement intensified utilization of regional arable land resources and land saving technology, and at the same time combine oriented nourishing technological system for arable land fertility with an aim to stabilizing arable land area and its quality.

It is to implement comprehensive management policies and measures of river basin water resources in case of climatic changes, and realize stable supply and reasonable utilization of river basin water resources.

It is to implement water-saving irrigation, minimal and zero tillage technology, farm straw transforming technology as well as high-efficient fertilization technological system. It is also to work on a series of highly efficient compound fertilizer and new controlled release fertilizer, as well as agricultural tillage and fertilization machines. It is to develop new crop species technology, to set up regulations of agriculture cultivation, to reduce the input of water, fertilizer and energy overall. Finally, they can save the cost of agricultural production.

It is to guarantee that medium and low yield land area should be reduced by 30% to 40%. The comprehensive utilization rate of soil, fertilizer and water of farm ecosystem should be improved by 10%. The nutrient and energy input should be decreased by 15% to 20 %. It is to apply compound fertilizer in grain crops widely, and to begin to apply slow released fertilizer for cash crops such as vegetables and fruits. It is to stabilize the quantity and quality of the arable land at national scale. It is also to set up three agricultural production technological systems, e.g., land saving agriculture, water saving agriculture, fertilizer and energy saving agriculture. They can lay the foundation for sustainable development of agriculture.

2. Mid-term goal in 2030

It is to realize the arable land substitution technology, high yield arable land conservation technology, medium and low yield land restoration technology, and to implement all-round management of constrained soil in realization of the stability and enhancement in terms of both quantity and quality of arable land.

It is to set up safeguard policies and technological system for river basin water resources, and secure the dynamic balance of water and soil resources on which the food safety production depends.

It is to work all-round on new multi-functional fertilizers for different regional climate and cultivation conditions, to set up new multi-functional fertilizer industry, to develop a series of slow/controlled released fertilizer and multi-functional fertilizer, and to do research on new technological system for biomass cultivation and transformation. It is to implement precision management of water and fertilizer and energy-saving cultivation integrating technological system. In coordination with the development of light energy utilization technology, it is to set up new and high-efficient planting system. It is also to set up water-, fertilizer- and energy-saving agricultural production system and enhance the utilization rate of agricultural resources.

It is to guarantee that medium and low yield land area should be reduced by 40% to 50%. The comprehensive utilization rate of soil, fertilizer and water of farm ecosystem should be improved by 20%. The nutrient and energy input should be decreased by 20% to 25%. It is to apply widely compound fertilizer and low-cost and slow released fertilizer for grain crops, and to begin to apply controlled released fertilizer and multifunctional biological fertilizer for cash crops including vegetables and fruits. It is to enhance the quality of arable land resources steadily and improve three major agricultural production technological systems of land saving agriculture, water saving agriculture, and fertilizer and energy saving agriculture. They can realize leapfrogging development of agricultural production in terms of intensification, mechanization, standardization, and industrialization.

3. Long-term goal in 2050

Based on the improvement of agricultural resource management information system, it is to set up resource security and regulation system for water and soil, with an aim to completely realizing the dynamic balance management of arable land and water resources.

It is to implement arable land nourishing technology, to control soil degradation processes in different agricultural zones, and to improve the arable land quality steadily.

It is to coordinate the development of new species of crops and material technology, and to develop manufacturing technologies of new intelligent fertilizer. It is to apply high-efficient utilization technological system of farm biomass resources. Based on the development of digitized and intelligent agricultural production technologies, it is to implement all-round precision management of water and soil suitable for crop growth as well as planting and tilling management system, realizing precisely mechanized agricultural production.

It is to guarantee that medium and low yield land area should be reduced by 50% to 60 %. The comprehensive utilization rate of soil, fertilizer and water of farm ecosystem should be improved by 30%. The nutrient and energy input should be decreased by 25% to 30 %. It is to apply widely slow released fertilizer and multi-functional biological fertilizer for grain crops, and to begin to apply intelligent fertilizer for cash crops including vegetables and fruits. It is to improve the quality of arable land resources and to establish all-round three major agricultural production technological systems of land saving agriculture, water saving agriculture, fertilizer and energy saving agriculture, and to stabilize the sustainable agriculture development.

5.3 A Roadmap to Technological Development

5.3.1 Key Scientific Tasks

Scientific tasks in the field of resource-saving agricultural technology can be defined in 7 aspects of 3 major fields. In the time- and land-saving agriculture field, the main tasks include: ① integrating key theories and technologies of arable land resources evolution, intensified management, and dynamic balance regulation and adjustment; and ② integrating key theories and technologies of degraded arable land prevention, restoration and oriented nourishing.

In the water-saving agriculture field, the main tasks include: ① Integrating key policies, technologies and systems of river basin water resources evolution and balance management; and ② water cycling mechanisms in soil-plant-air continuum (SPAC) and key theoretical technologies and relative machinery for water saving agriculture ecosystem.

In the fertilizer and energy-saving agriculture field, the main tasks include : ① material science and key production technologies of slow/controlled released fertilizer and intelligent fertilizer; ② recycling and efficient utilization of nutrient resources from farm ecosystem, and integrating key theories and technologies of coupled water and fertilizer; and ③ key theoretical technologies of minimal and zero tillage and relevant machinery.

5.3.2 Design Philosophy for the Roadmap

According to the adjustment of agricultural structure, the national conditions and the requirement of agricultural functions development, making use of multi-disciplinary means of resource management science, ecosystem ecology, soil science, agricultural machinery, material science and information science, and taking resource saving agriculture and production capacity of arable land as an important scientific task, three aspects of theory and technology systems will be set up including arable land protection and improvement, efficient water resources utilization, nutrients and energy saving and use. In light of the development of agriculture information system and crop species three steps will be implemented: ① by the year of 2020 it will be to stabilize and balance the supply of water and soil resources, primarily reduce the consumption of water, fertilizer and energy resources, and stabilize the production capacity of arable land; ② by the year of 2030 it will be to realize intensified and standardized management of water and soil resources, overall reduce the consumption of water, fertilizer and energy resources, and overall enhance the production capacity of arable land; ③ by the year of 2050 it will be to overall realize the digitized and dynamic management of water and soil resources, as well as the precision management of water, fertilizer and energy resources, and stabilize the high production capacity of arable land.

Firstly, it is to set up monitor system for arable land, to establish the platform for water, fertilizer and energy utilization, and to build the platform for new fertilizer research and development. Secondly, on the basis of the theories of evolution and adjustment of regional arable land and water resources, cycling and controlling of water, fertilizer and energy of farm ecosystem, prevention, restoration as well as oriented nourishing fertility of degraded soil, it is to realize intensified utilization of water and soil resources, to achieve breakthrough in land saving technology, water-saving irrigation technology, minimal and zero tillage technology, high-efficient utilization technology of farm biomass resources, as well as high-efficient fertilizing technological system, to research and develop matchable agriculture machine, and to formulate standardized production regulations. Thirdly, it is to develop new types of slow and controlled released fertilizer and intelligent fertilizer, and to set up new fertilizer industry. Finally, in light of the development of agriculture information system and crop species, it is to establish all-round three principal production technological systems of time- and land-saving agriculture, water-saving agriculture, and energy-saving agriculture, and to realize intensified production of agriculture and stable improvement of arable land production capacity.

Table 5-1 Roadmap of Resource Saving Agriculture Development

	Phase	2020	2030	2050
Phase target	Target	① Establish monitore and forecast platform for arable land and water resources, platform for water and fertilizer utilization research, and platform for new-type fertilizer research and development ② Establish arable land coordinating and coupling utilization and soil quality directive breeding technologies ③ Research on water-saving irrigation engineering technology and its integrating soil tillage, water saving and straw cover technology ④ Research on fertilizer saving and zero tillage technologies as well as new slow/controlled released fertilizer	① Establish arable land substitution technology and constrained soil restoration technology ② Research on various water saving technology and management of river basin water resources ③ Research on mechanized technology of water-fertilizer-energy coupling management and multi-functional biological compound fertilizer	① Establish large-scale operation and management system of regional arable land and high-yield soil nourishing technology ② Establish all-round and high-efficient utilization system of river basin water resources and engineering-biology-chemistry water-saving integrated technology ③ Establish comprehensive management system of regional agricultural nutrient resources and develop intelligent fertilizer
	Indicator	① Medium and low yield land area down by 30-40% ② Utilization rate of water and fertilizer up by 10% ③ Nutrient resources input down by 15-20% ④ Apply compound fertilizer and slow released fertilizer widely ⑤ Stabilize quantity and quality of arable land resources	① Medium and low yield land area down by 40-50% ② Utilization rate and water and fertilizer up by 20% ③ Nutrient resources input down by 20-25% ④ Apply slow/ controlled released fertilizer and multi-functional fertilizer widely ⑤ Enhance the quality of arable land resources steadily	① Medium and low yield land area down by 50-60% ② Utilization rate of water and fertilizer up by 30% ③ Nutrient resources input down by 25-30% ④ Apply intelligent fertilizer widely ⑤ Realize the improvement of quality of arable land resources
	Realization of intensified-, mechanized-, standardized- and industrialized- agriculture production			

Continued

Scientific tasks	Land saving agriculture technological system	Monitoring and forecasting platform for the national arable land
		→ Arable land coordinating and coupling utilization technology
		→ Arable land substitution technology
		→ Large-scale operation and management system of regional arable land
		→ Prevention of soil degradation and oriented soil nourishing technology
		→ Degraded soil (constrained) restoration technology
		→ High-yield soil nourishing technology
	Water saving agriculture technological system	Monitor and disaster forecast platform for water resources
		→ Integrated technologies of soil-tilling, soil moisture-preserving and straw-covering and surface water-saving irrigation
		→ Species water-saving technology and limited irrigation technology
		→ Integrated system of water and fertilizer and engineering-biology-chemistry water-saving technology
		→ Joint allocation technology of surface water and ground water at watershed scale
		→ Water resources knowledge management and technological system at watershed scale
		→ Water-saving society construction and management at watershed scale
	Fertilizer and energy saving agriculture technological system	Farm nutrients cycling research platform and new fertilizer research platform
		→ Compound fertilizer, organic fertilizer and biomass resources multi-level transformation technology
		→ New slow/controlled released fertilizer and multi-functional biological compound fertilizer
		→ Intelligent fertilizer
		→ Balance between fertilizing, minimal and zero tillage and planting system
		→ Precision mechanized application technology of integrated management of water, fertilizer and energy
		→ Regional farm nutrient resources management system and three dimensional agriculture ecosystem

5.3.3 General Roadmap

1. Technological system for land saving agriculture production and arable land production capacity improvement

Phase 1 (2010–2020): It is firstly to establish research platform for arable land monitor and management technology at regional scale on the basis of intelligent wireless network monitoring technology and distributed data collecting technology. It is to develop arable land protection monitor and forecast technology, and do research on coordinating utilization technology and planning technology for regional arable land. Meanwhile, It is to work on prevention technology of various types of soil degradation (erosion, acidification, nutrients depletion, non-equalization, biological function degeneration, desertification, salinization and alkalization) as well as oriented nourishing soil fertility technology for medium- and low- yield arable land, in order to stabilize the quantity and quality of regional arable land, and to stop the downward tendency of quantity and quality of regional arable land.

Phase 2 (2020–2030): It is to develop different regional arable land substitution technology, e.g., arable land replenishment, arable land resources exploitation, and arable land reclamation. Meanwhile, it is to work on comprehensive restoration technology of degraded soil in terms of chemistry, physics and biology (also including restoration agent), bring constrained soil under complete control and improve medium- and low- yield arable land via soil fertility oriented nourishing technology in order to achieve stable supply and appropriate use of arable land resources and to enhance the quality of arable land.

Phase 3 (2030–2050): In light of improved agriculture information management system, it is to establish digitized integrated management system for arable land resources as well as to propose regional policies and measures for large-scale management of arable land. Meanwhile, it is to develop and adopt technology of arable land saving system, and to enhance the quantity and quality of regional arable land resources steadily, and to maintain the dynamic balance of the essential arable land resources on which the security of national grain production depends.

2. Water saving agriculture ecology technological system

Phase 1 (2010–2020): In coordination with research platform for nutrients cycling utilization of agriculture ecosystem, it is to establish management platform for water resources and research platform for tillage technology. It is to work on integrated exploration and utilization technology for surface water and ground water at watershed scale on the basis of ET management in order to realize water balance in watershed agriculture. It is to work on engineering and chemical technologies for water saving as well as integrated technologies for soil tilling, soil moisture saving, straw covering and surface water saving irrigation. Meanwhile, it is to do research on water saving machines in order to save water resources and reduce the cost of agriculture production.

Phase 2 (2020–2030): It is to work on anti-drought species and water and fertilizer coupling technology, and to integrate comprehensively biological water saving and limited irrigation technological system, and to establish water-saving farm ecosystem construction technology on the basis of knowledge management (KM) at watershed scale, to establish water and fertilizer precision management system at watershed scale, to establish overall water-saving production technology and enhance the utilization rate of water resources in the farm field.

Phase 3 (2030–2050): It is to implement the integrated system of water and fertilizer coupling and engineering-biology-chemistry water saving technology at watershed scale, to build overall management system of watershed water resources, to implement security policies and technological system of watershed water resources, to construct water saving society, and to maintain the dynamic balance of the essential water resources on which the security of the national agricultural production depends in case of climate change, in order to realize stable supply and reasonable utilization of water resources at watershed scale.

3. Energy saving agriculture production technological system and high-efficient utilization of nutrients technological system

Phase 1 (2010–2020): Firstly, it is to set up research platform for nutrients cycling utilization technology in farm field, to establish research and development platform for the new-type high-efficient fertilizer, and to establish technological center for fertilizer engineering. In coordination with the development of crop species technology, it is to work on high-yield and super-high-yield planting modes and high-efficient fertilizing agriculture technological system, and to develop matchable mechanization technology for whole production process. It is also to work on high-efficient utilization technology of renewable biomass resources, e.g., farm straw burning control, agronomy substitution technology. It is to develop high-efficient compound fertilizer, special fertilizer and new-type slow/controlled released fertilizer technology and their industries. It is to do research on minimal and zero tillage technology and energy saving technology. It is to apply high-efficient compound fertilizer widely in order to enhance utilization rate of nutrients in agriculture ecosystem and to lower nutrients and energy input.

Phase 2 (2020–2030): It is to constitute overall technical regulation for high-efficient agriculture productions. In coordination with the development of crop species and water saving technology, it is to develop mechanized precision application technology for water-fertilizer-energy integrated management and supporting facilities. On the basis of compound fertilizer and slow/controlled released fertilizer manufacturing techniques, it is to develop new-type and multi-functional biological compound fertilizer in light of different climatic conditions and cultivating conditions. It is to establish new-type and multi-functional biological compound fertilizer industry with independent intellectual property rights and popularize slow/controlled released fertilizer and multi-functional biological fertilizer, with an aim to enhancing the popularization rate of multi-functional biological fertilizer and the utilization rate of nutrient in farm land.

Phase 3 (2030–2050): In coordination with the development of new crop species and new material technologies, it is to develop manufacturing technologies for new intelligent fertilizer and comprehensively apply intelligent fertilizer. Availing of the information-based and intelligent-based agriculture it is to overall achieve water and fertilizer precision management and coupled cultivation and tillage management technological system which are best suitable for crop growth. It is to overall establish comprehensive management system for regional agricultural nutrients resources and stereo agriculture ecosystem in order to achieve high- and stable- yield, high-efficient and high-quality agriculture production system.

Main References

[1] Cao ZH, Zhou JM. Soil Quaility of China. Beijing: Science Press, 2008.

[2] Qian ZY, Zhang GD. Comprehensive and Dissertation Report of Water Resources Strategy for Chinese Sustainable Development. Beijing: China Water Conservancy and Hydropower Press. 2001.

[3] Xia J, Huang GH, Pang JW, et al. Sustainable Water Resources Management: Theory, Approach & Applications. Beijing: Environmental Sciences & Engineering Publication Center, China Chemistry Industry Press. 2005.

[4] Wang H. Chinas Water Resources and Sustainable Development. Beijing: Science Press, 2007

[5] Ren GY, Jiang T, Li WJ, et al. An integrated assessment of climate change impacts on China's water resources. Advances in Water Science, 2008, 19(6): 772-779.

[6] Society Department of the China Association of Science and Technology. Research on Land Degradation in China: Proceedings of National Conference on Land Degradation Research. Beijing: China Science and Techonology Press. 1990.

[7] Zhang XL, Gong ZT. Human-induced soil degradation in China. Ecology and Environment, 2003, 12(3): 317-321.

[8] Zhu ZL, Norse D, Sun B. Policy for reducing non-point pollution from crop production in China. Beijing: China Environmental Science Press, 2006.

[9] Hydrology Bureau of Ministry of Water Resources, P R China. Water resources quality assessment for China. Beijing: China Science and Techonology Press. 1997.

[10] Chen Y, Zhang QH, Zhang HF. Suggestion of utilization and protection of plowland resource in China. Resources & Industries, 2008, 10(5): 4-8.

[11] Zhao BD; Zhao QL. A preliminary study on the resources-saving agriculture production systems. Journal of Henan University (Natural Science), 2001, 31(4): 66-69.

[12] Pennisi E. Plant genetics: the blue revolution, drop by drop, gene by gene. Science, 2008, 320: 171-173.

[13] Finkel E. Richard Richards profile: Making every drop count in the buildup to a blue revolution. Science, 2009, 323: 1004-1005.

[14] Wu ZJ, Zhou JM. Develop release reduced/release controlled fertilizer, increase fertilizer efficiency. Bulletin of the Chinese Academy of Sciences, 1999, 5: 356-360.

[15] Wu ZJ, Chen LJ. Release Reduced/Release Controlled Fertilizer: Theory and Application. Beijing: Sciene Press. 2003.

[16] Shaviv A. Advances in controlled release of fertilizers. Advances in Agronomy, 2000, 71: 1-49.

[17] Fan XL, Liu F, Liao ZY, et al. The status and outlook for the study of controlled-release fertilizers in China. Plant Nutrition and Fertilizer Science, 2009, 15(2): 463-473.

[18] Kintisch E. Renewable energy: Minnesota ecologist pushes Prairie Biofuels. Science, 2008, 322: 1044-1045.

[19] Fischer JR, Finnell JA, Lavoie BD. Renewable energy in agriculture: back to the future. *In*: Duffield JA. Choices, Biofuels: Developing New Energy Sources from Agriculture. The Magazine of Food, Farm, and Resource Issues, 2006, 1: 27-31.

[20] Cao WG, Duan H. Present situation of utilization and resources of biomass energy in China. Journal of Anhui Agricultural Sciences. 2008, 36, 14: 6001-6003.

[21] Yang LZ, Sun B. Cycling, balancend management of nutrient in Agro-ecosystems in China. Beijing : Science Press. 2008.

6 Roadmap of Agricultural Production and Food Safety Science and Technology Development

As Food is related to human survival and health, the country's reputation and image, and the economic benefit and position in the international trade, it has become one of the hottest issues in the world now. As the continuous growth of world population, the constant changes of human life style and the continued deterioration of natural environment, higher requests for food safety are put forward by human society. Therefore, we should not only ensure the safety of quantity, but also to ensure the safety of quality in the food safety field by 2050. New concepts and technological means should be adopted to ensure the constant production and supply of the "green" and safe agricultural production to meet the people's great need for food safety, nutrition and health.

6.1 The Demand and Significance of Development

"The people take food as their prime want", food is not only the most basic human survival needs, but also the eternal theme for the national stability and social development. The food industry in 21st century is facing tremendous changes, and shows two important trends: The first is to improve the food safety; the second is to enhance the food nutritional value.

6.1.1 Address the Basic Needs of Food Safety

Food safety is the basic needs for the development of human society. Broadly speaking, the food safety includes the quantity safety, quality safety and nutrition safety. With the life sciences and related disciplines on the enormous support of agricultural production, grain output in China has increased exponentially, and the problem about food and clothing has been basically solved. Therefore, from the social development and current situation, food security means more about the food quality safety.

1. Eliminate the hazards that influence the food safety

As we all know, with the continuous improvement of living standards, people's requirement on food has changed from the "quantity" to "quality". However, because of the restriction of the level of productivity, management capability and technology development as well as other factors, there are still many alarming issues in the food safety field. In the past 10 years, the incidence of foodborne diseases worldwide presented an ascending trend [1-3], with about 4–6 billion cases each year. There are about 180 million people died from such diseases in the developing countries [4,8]. Even in the developed countries, more than 10% of people were infected with foodborne diseases each year, which not only seriously harms people's health, but also causes enormous economic losses.

(1) Excessive use of pesticides and residues

The world's annual production of chemical pesticides was up to 200 tons, half of which was consumed in China. However, only about 35% of pesticides were absorbed and used; most of residue stay in the fields or flows into surface water and groundwater, causing environmental pollution. With agricultural products such as rice, vegetables, fruits and tea containing excessive levels of pesticide residues through using chemical pesticides, the food security problems are getting more and more serious and have become the focus that national and public pay close attention to these issues.

(2) Hormones and veterinary drug residue and pathogen pollution

With the development of industrial and intensive animal breeding, growth hormone, preservatives, appetite enhancer, clenbuterol, artificial colors and other feed additives and antibiotics, sedatives and other veterinary drugs have been used in animal production. The abuse of additives and drug has led disease resistance of pathogenic microorganisms and drug residues of animal to reach alarming proportions. At the same time, the early warning system for animal disease prevention and control is also facing the grim challenge of the intensive and large-scale farming model. The animal food pathogen contamination [5-7, 10] (pathogens, viruses and parasites that people and animals suffered from altogether) caused by animal diseases not only threatens human life and health [11-12] and brings the fatal blow to the livestock breeding, but also will affect social stability, and cause public panic. The outbreaks of bird flu [9, 14] and Streptococcus suis [15-18] in recent years are the typical instances.

(3) Heavy metal pollution and accumulation

Due to affection of sewage irrigation, industrial activities and organic agricultural fertilizer and other factors, the heavy metal pollution in the environment has imposed tremendous pressure on the safety of agricultural products. The main heavy metal pollution in our country's agricultural products include Hg, As, Pb, Cd, Cr, Cu, Cu, Zn, etc. Long-term consumption of Cd and other heavy metal has caused high incidence of cancer among farmers in Shangba Village, Guangdong Province; there are other cases as to vegetable with high level of Pb and Cd in some areas of south Jiangsu province. In addition, the chemical additives used in the process of livestock and aquaculture also lead to heavy metals accumulated in animal liver, kidney or other organs. Thus, it is extremely urgent to control the heavy metals pollution in the agricultural production and processing.

(4) The dangers of biological toxins

Biological toxins refer to toxic substances produced by a variety of organisms (animals, plants, microorganisms). Biological toxins come in a wide variety and some of them such as the aflatoxin and Aspergillus varsicolor toxins in maize and peanut have been proved to be the main inducers for the regional liver, stomach and esophagus; the tetrodotoxin, diarrhetic shellfish poisoning, paralytic shellfish poisoning and neurotoxic shellfish toxins in the aquatic products can have enormous influence on physiological functions of the human body. Mycotoxins are secondary metabolites produced by fungi and more than 200 kinds of them have been known at present. The intake of agricultural and livestock products contaminated by mycotoxin may make human or animals suffer from food poisoning. More seriously, many fungal toxins that have carcinogenic, teratogenic and mutagenic function will cause chronic damage to the human body. The treatment and pollution control of biological toxin poisoning is still a worldwide problem today.

2. Create the standard technology system of safe agriculture products

(1) Build the standard production model of planting

Agricultural standardization refers to the use of the standard principles of "unification, simplification, coordination and preference" depending on the advanced agricultural science and technology achievements and experience to make agricultural produce antenatal, medium and postpartum processed standardly, to ensure the quality and safety of agricultural products. In developed countries, the agricultural products basically have achieved the standardization, and have established a relatively complete support system of agricultural standardization. At present, the green food standardization is the most influential one. For the green food production, firstly, it emphasizes the best ecological environment; secondly, it emphasizes the full-process production management. It proves that the agricultural standardization is playing an increasingly important role in the quality and safety of agricultural production.

(2) Build the standard and safe (green) aquaculture production systems

In the process of animal production, we should create an excellent breeding environment, select and introduce the higher-yield varieties with strong resistance to diseases not carrying pathogens. Also it is necessary to take the precaution of animal disease, improve the prevention and control system, cut off the spread of disease, implement the usage standards of veterinary drugs and feed additives strictly, add the growth, immune modulators with no residual and non-toxic side effects and anti-stress additive to promote and control animal deceases as much as possible. The establishment of animal production of standardized green farming system is an effective way to guarantee the good quality and safety, no harmful residue and no pathogen pollution in animal food.

(3) The forecasting of pest and build of prevention and control system

The excessive use of chemical pesticides has led to increasingly shorter cycle of pests and increases of pest types leading to a vicious cycle of agricultural production. It has proved that using different types of natural enemies and biopesticides to create whole biocontrol system is the fundamental way to ensure the food safety and reduce use of chemical pesticides.

Pathogens resistance to drug and animal products with ultra-high dose of drug residues is another important factor to threaten animal product safety. The purification of pathogenic microorganisms in the environment and animals, improvement of the animal disease early warning and prevention and control system and the research and development of new green veterinary drugs can reduce animal disease and pathogens and drug residues in animal products and establish standard livestock production systems.

(4) The agricultural postpartum storage and processing system

The fruit and vegetable production were 71 million tons and 1.4 million tons, respectively in our country now, and their output will reach 240 million tons and 350 million tons, respectively, by 2050. On one hand, postpartum rot is very serious and have caused the post-harvest losses up to 30 billion yuan each year. On the other hand, because of their special nutrition and flavor, the processed foods with fruits and vegetables as the main raw materials are very popular. Therefore, based on the use of unique plant, fruit and vegetable as raw materials, it is important to develop the innovative processing technology to produce new functional and specialty foods for the future agriculture and industrial development.

(5) Agricultural products safety monitoring and evaluation system

Food safety inspection and testing is one of the most important means to monitor food safety. There is still a wide gap in the development of inspection technology in China compared with international level, so we now could not track and study the hot issues of international food quality and safety. The cases of "Sudan", "Melamine" and "Clenbuterol" reflected the method and level of our country's food safety inspection falls behind and the emergency testing capability is so weak that it is difficult to meet the requirements of new situation.

The long-term and short-term risk assessment on hazardous substances of various foods is a good countermeasure for early-warming. Countries in the world have increasingly established the food safety control technology system that is core with risk assessment and integrates assessment, detection, monitoring, early warning and control as a whole, while China has just started in this area. Therefore, it is of great significance to carry out the study of food safety risk assessment and create the technology system of food traceability and early warning.

3. The secure environmental quality is a necessary condition for food safety

At present, due to the effect of the high-intensity use of agricultural resources and the rapid development of industrialization and other factors, the soil, water and air in China have been badly polluted, and the agro-ecological has been seriously threatened. The degradation of agricultural environment not only enhanced the investment in agriculture and reduce the output, but also increased the polluting extent on agricultural production. Pollutants can affect agricultural products safety, threaten human health and affect international agricultural trade in many ways, such as crop root and leaf absorption, aquatic animal consumption and food chain enlargement. With the further improvement of the agricultural products quality standard of international market, the agricultural products in our country will suffer from more "green barriers" blockade and face more severe challenges.

6.1.2 The Diversity of Needs for Research on Nutrition

As people are concerned more about health and the demand for food diversity and quality, the nutrition research and development has aroused greater attention as people's living standards improved.

1. Value the nutrition quality is an objective requirement of the food

Nutrients are the material foundation for the survival and development of humanities. Food is the carrier of nutrients and nutritional function is the natural attributes and basic function of food. However, the nutritional quality of food could be reduced by the food processing and result in the loss of food nutrition. Therefore, as an important part of the food industry, there is a natural and necessary link between the agricultural production and processing and the nutritional quality. The nutrition science and philosophy have been generally applied in western developed countries to guide the development of agricultural production and processing industry, to optimize the industry structure, to apply the advanced technology purposefully, and better meet people's nutritional health needs and market demands.

2. Improvement of living standards demands the improvement of the nutritional quality of food

After 30 years' rapid economic development, society of China has achieved well-to-do level and the Per capita GDP has reached 1,000 U.S. dollars. International experience tells us that after the GDP per capita national income reached 800 U.S. dollars, the food consumption will enter the rational stage, the diet will change, and people will make better food quality requirements and want to consume foods with more nourishment and the role of disease prevention and health care. The ideas of eating nutria nutrient and healthy food have been accepted by more and more people. It is predicted that the proportion of nutrition and health consumption in the whole food consumption will become heavier.

3. Pay attention to the food nutrition for people nutrition and health

According to the National Nutrition Survey reports published in 2002, the residents of our country was facing the dual challenges of nutritional deficiencies and structural imbalances, especially some micronutrients such as calcium, iron, iodine and VA are seriously deficient. The calcium intake of the people in our country is less than 50% of the standard recommended amount. There are 5.8 billion people in marginal VA deficiency states; 200 million people suffer from iron deficiency anemia and 1.6 billion people lack of energy and protein intake. At the same time, the nutrition-related diseases caused by imbalance nutrition intake show a high incidence trend. There are 200 million people overweight, more than 6,000 people obese, 40 million people with blood sugar unusual, and the number of people suffer from hypertension or dyslipidemia is 1.6 million. If the sub-health population affected is taken into account, the number will increase by several times. The rate of overweight adults increased by 39% and obesity increased by 97% compared with the National Nutrition Survey data in 1992, which shows that the nutritional status of Chinese residents has not been improved synchronously along with the rapid development of economic, but presents the characteristics of both complex structure and total enlargement.

The severity of the nutritional status of our country is closely related to the unilateral development of the agricultural and food production. Over the years, much more attention has been paid to the quantity increase and technical improvement in the agricultural production and processing, but less emphasis on the nutrition quality of products and production of nutrition foods. So far, we have not integrated the modern nutritional science into the principles to guide the development of production and processing of agricultural products, and have not fully established and popularized the knowledge and philosophy of modern nutritional science.

The western countries had begun to implement the student nourishment and develop the nutrition programs since the 60's of 20th century, which not only significantly improved the nutrition and health of students and residents, but also effectively stimulated the agricultural structural adjustment and promoted the rapid development of the food industry. From the present Chinese agricultural products technical level, there is a high proportion of primary processing of raw materials while the deep processing level of raw materials and the technological content of manufactured goods are low. Therefore, to integrate the philosophy of nutritional science and the production of nutritional foods into the production and processing of agricultural products is not only the common experience of the developed counties of the world, but also the necessary choice to fit in with the people's nutritional health requirements.

6.1.3 The Development of the Individual Needs of Multi-functional Food

The green foods and functional foods will become an important part of modern life. With the social development, the living standards of people continuously improved, but some diet-related diseases, such as the iron deficiency anemia, hypertension, diabetes, high blood cholesterol, reduced immunity, coronary artery and senile dementia, have increased dramatically, and have become a social problem that seriously influenced the health and well-being of people. According to a survey released by World Health Organization (WHO), the proportion of the sub-health has accounted for the 75% of the population of the world. The public concerned most is weight control; enhance immunity, anti-oxidants and nutritional supplements. In the United States, nearly 60% of people taking the nutritional supplements that contain a variety of vitamins and minerals.

Functional foods, with many characteristics, such as with no dose limit, it is safe to consume under normal conditions, with nutritional and healthy functions and have defined consumptive objects and so on, have been acknowledged as the 21st Century Foods. According to the special physical conditions of the consumer groups, the special functional foods could be developed by scientists. The significance is to emphasize the regulatory function of food on the disease prevention, rehabilitation and health improvement. At present, the foreign countries begin to focus on the research and development of functional foods. Its main functions are: regulation of physiological activity rhythm; regulation and enhancement of body's immune system; regulation of the state of mind, extending aging and maintaining the vitality of the human body, etc.

1. Nutrition balance food for different groups

(1) Target for children, the elderly and women

The irregular daily life and diet lead to a low-aging tendency of chronic diseases, more and more young people and children show high blood fat, overweight, high blood pressure, unbalanced nutrition, fatigue and other symptoms. For example, there are 7.2 million children in the United States, including 270 thousands of children under the age of 19 suffering from high blood fat, 2,000 thousands people under the age of 26 having high blood pressure, 60% of the children feeling tired because of lack of energy and 15% of school children falling asleep. The functional foods which are suitable for children, such as enhanced convenience breakfast foods and dairy products rich in active bacteria are very popular. The vitamin, DHA, EPA and β-carotene are also popular with teenagers and children alike.

As public's expectation for longevity increase, the functional foods have attracted the consumer's attention that can protect brain, help to lose weight, increase appetite, promote digestion, benefit the skin and black the hair, raise eyesight, strengthen immunity, prevent dementia and improve memory.

Women are the target objects of the functional food industries. There are many food items designed specifically for women in the market, even including the nutritional drink and bread containing isoflavones that developed specifically for postmenopausal women. Studies showed that the soy isoflavones could effectively control the menopausal syndrome, osteoporosis, breast cancer, and had the antioxidant and cholesterol lowering physical activity and the role in beauty.

(2) Target for the sub-health adult population

Sub-health and chronic diseases are common in the modern fast-paced life and their harm to human body is undoubtedly enormous. According to statistics, there are 600 million people between 20-45 years old suspected under sub-health condition in the United States. In China, the number of the people under sub-health condition is more than 700 million, accounting for the 60% to 70% of the total population. More and more adults buy nutrient supplements to improve the side effects by the nutritional imbalance and fatigue. In the United States, 29% and 36% of males and females, respectively, concern about mental health, and the brain energy products also appear on the market. Three-quarters of Americans need the food that can enhance the immune system. The products made from lutein, anthocyanins and carotenoids which can improve eyesight have appeared on the European and U. S. markets.

Obesity has become a serious public health problem in the world. There are nearly 105 million adults over 20 years old overweight and 425 million obese in the United States. Therefore, the functional foods including low-calorie diet food is extremely popular in the United States.

2. The personalized foods that target for people with different physiological and health status

(1) Take public health as the goal

With the popularity of the public knowledge of nutrition, "make the functional foods popular, make popular foods functional" will also be the development trend of functional foods. More and more consumers purchase functional foods for health purpose, and for health, the public concerned most including weight control, enhance immunity, anti-oxidation, nutrient balance, etc. Some active ingredient extracted from the natural plants and animals, such as the lycopene, lutein and other antioxidants, has been paid close attention and developed very fast.

(2) Improve body health and mental state as the goal

Among the functional foods that can provide energy, the sports nutritional foods and drinks are the most popular. For example, the functional tea is available anywhere in the market at present. Some functional foods targeted to increase "brain power", anti-allergic, anti-stress, improve eyesight also appeared on the market. There are still many other functional materials which are very popular among the consumers, such as allicin, peptides, soy isoflavones, chitosan, ω-3-polyunsaturated fatty acids and whey protein and so on.

(3) Reduce the chronic disease risk as the goal

Use functional foods as the auxiliary medical treatment to reduce the symptoms and risk of the disease will become a main channel for the development of functional foods. There are 970 million people worldwide suffering from high blood lipids and 14.3 billion people overweight. About half of Americans believe that the risk of disease could be reduced by using some food instead of drugs. In addition to using the functional food to reduce risk of cardiovascular disease, cancer, obesity and diabetes, consumers also purchase some functional foods with anti-allergic effect and other healthy food to alleviate osteoporosis, promote gastrointestinal health, prevent dental caries and improve joint pain. For example, the probiotics can improve the function of gastrointestinal tract and reduce the risk of stomach disease.

6.2 Status and Trend of Technology Development

The implementation of sustainable agricultural development strategies and the development of "Green Agriculture" will be the developmental direction of the agricultural economic for every country in the world. Currently, the world economic development and consumer market is undergoing profound changes; the natural and pollution-free products have become a kind of new consumption fashion. The future agricultural production must be in clean soil and produced with clean manner to produce safe food (green food) to meet the global food consumption in the quantity and quality needs. China's Ministry of Agriculture officially launched the Pollution-free Food Action Plan in 2001. This plan focused on the control of pollution at resource, vigorously promote standardization of agricultural production, and comprehensively improve the quality and safety of agricultural products. The research and demonstration of the green and safety production techniques have been greatly improved. From the controlling technique of the quality of producing area conditions, the green controlling technique of production process, the high efficient utilization technique of recycling resources, quality evaluation technology to the establishment of the standardization of quality control and production techniques of agricultural products have been greatly developed and demonstrated. "The People's Republic of China Agricultural Product Quality Safety" has been come into force on November 1, 2006, and the safe production of agricultural products will be standardized from the legal point of view.

6.2.1 The Development and Trend of Safe Food

As foodstuff has the function of food and health care, the safety issues are causing more and more attention. As early as 1970's and 1980's, developed countries and relevant international organizations have developed the residue standards of food pesticides, heavy metals, nitrates, veterinary drugs and antibiotics. At present, the main trends of the international development of safe food are expressed as the following aspects: attention to the cultivation and development of green food, including organic food, natural food, green agricultural products; focus on the research and develop the alternative chemicals, bio-fertilizer and pesticide; make environment and health as the preferential domain to develop, emphasize to reduce pesticide and fertilizer input, and improve the chemical substances control system to reduce the risk of chemical substances.

1. Plant pests control technology

In recent years, the EPA has withdrew the registration of 59 chemical pesticides; EU has set the strategic plan to reduce chemical pesticide use; China also has banned 40 kinds of chemical pesticides used in vegetables and listed "green food and bio-pesticides" as one of the priority development projects into "21st century agenda of Chinese". The research and development of biological control is not only an important part of the modern agriculture, but also the focus of hi-tech competition of international agriculture. Research and development of new pesticides (bio-pesticides or green chemical pesticides) has become the mainstream of the development of modern agriculture. The research of new pesticide is focusing on the synthesis and use of insect pheromones, microbial biological control agents, biological antagonistic bacteria, plant stress inducer, fungicides, insecticides and herbicides, insect virus pesticides, fungi pesticides, and botanical pesticides based on the research of microbial protein.

The main products include microbial pesticides (such as pesticides, insect viruses), anti-inducer (Messenger, abscisic acid, oligosaccharides), biochemical agents (such as the sterilant rodents, insects pheromones, etc.), biological control predators (such as *Trichogramma*), and botanical pesticides (such as matrine, Pyrethrins, etc.). Since 1990s, the global production of biological pesticides is annual increase with the speed of 10-20%. There are about 30 kinds of bio-pesticides in the international market. Including more than 10 biotechnology products, the production value exceeded one billion U.S. dollars.

While China has taken various measures to carry out the prediction, prevention and control of agricultural pests and diseases and study the establishment of GIS-based decision support system for the crop pest management and the prediction model of agricultural pests and diseases based on the artificial neural network, there is still a big gap between China and the advanced countries on the full application of biological control technology to control pests and diseases. We still cannot be like the Nordic and North American to provide the producers with biocontrol technology package, like various natural enemies and more varieties of bio-pesticides, to control pests in some complicated farm land and green house environment.

2. The development of plant resistance inducing factor

The inducible resistance system existing in the plants is similar to that of animals. Many functional materials that can induce plant resistance have been discovered by domestic and foreign scientists and applied to prevent and control plant diseases. During the long-term process of evolution, plants have developed some capabilities and characteristics, including the plants' own immune system and the resistance capabilities that induced by the external factors or elicitors, which can protect themselves from the invasion of pathogens. When stimulated by the inducing factors of external environment, plants can resist to the disease. The factors that can induce or stimulate plant resistance include the virus capsid protein, antagonistic bacteria, germ weak lines, ABA (S-induced resistance factor), oligosaccharide, and elicitor protein. When these inducing factors or elicitors contact with plants and act on plant tissues, ethylene, salicylic acetic acid, jasmonic acid phytoalexin and pathogenesis-related protein can be produced through the signal transduction. These materials can improve and enhance the ability of plant resistance to pathogen invasion, prevent the occurrence of disease or reduce the disease extent by means of the regulation of plant metabolism and activation of plant immune system and growing system to, so as to reduce the use of pesticides radically, and alleviate the environment and agricultural products pollution caused by pesticides at the source.

3. Animal health breeding

In order to meet the requirements of technical safety, high quality and effective production of animal products, the metabolism and regulation of animal nutrition, animal environmental control and feeding techniques, the research on animal waste harmless and value-added processing, the disease control in the livestock and poultry breeding process and the establishment of standard of health culture has been the core content of the international animal research since the 90' of 20th century. At present, the research on the key technology of health culture and the standard-setting has become the most direct and effective means to implement green technical barriers of animal culture in countries around the world. The over control of "from farm to fork" is the necessary means to protect the safety of animal food, and for this reason the HACCP, GAP and other management systems have been developed. Based on the highly consideration of animal products and feed safety, WTO member states in accordance with the WTO-TBT/SPS (Technical Barriers to Trade / Animal and Plant Quarantine) agreement have developed the laws, regulations and standards for animal products trade. How to maintain the animal product safety, high quality and efficient production and realize the livestock breeding health development has more than their own sustainable development issues, also related to international relations in trade, politics and even national security issues. Under this background, countries around the world competing to carry out the research on animal welfare, animal food safety, high quality and efficient production, to get more favorable position in future international market contests.

4. The safe aquiculture breeding technology

At the beginning of the new century, some aquaculture developed countries have regarded the sustainable development of aquaculture, especially in food security and ecological security as part of national development strategies to re-validation, and worked out the aquaculture development and technology planning. For instance, Japan in March 2002 developed a comprehensive policy plan based on the guiding philosophy of from the "Basic Plan for Fisheries" to "ensure stability of supply of fish" and "the sound development of aquaculture". National Oceanic and Atmospheric Administration Fisheries Board developed aquaculture technology strategy, the

EU released a new common green paper on aquaculture policy, aims to establish a responsible and sustainable aquaculture.

Aquaculture in our country speedily developed, benefiting from the development of fishery science and technology and the big breakthrough in technology. For example, the artificial reproduction technology of "four everybody fish", artificial breeding techniques of seaweed, laver, scallop and industrialization of prawn, abalone hybrid technology, the theory and technique system of comprehensive high-yield fish culture in large ponds, large water surface " triple nets" (cage, pen, mesh) for proliferation of fish and integrated support fish culture technology for resources enhancement and fertilizer, prevention and control technology for outbreak of epidemic, modern industrial fishing technology, which provided technical support for the rapid development of aquaculture in China.

5. Pay attention to development of recycling economy

Since the strategies for sustainable development has been put forward in 90's, developed countries have been regarding the development of recycling economy and establishment of recycling society as an important strategy and way to implement the sustainable development strategy. In 1996, a new environmental law-"Circular Economy Law" came into effect in Germany. The core idea of this law is based on reduce, reuse and recycle 3R principle to keep more materials in the production circle. From the viewpoint of domestic or oversea research status, the research on the fluxes process and environmental effects of water, fertilizer and drug in the farmland ecosystem of domestic is still weak, the quantitative research is not enough; the regional spatial nitrate leaching index assessment has not carried out; the development and application in agricultural producing safety warning system also weak. The research on safety production of agricultural products is more from a technical perspective, and is inadequate to carry out system optimization management model from the overall health of the system functionalities. The application of efficient and precise technology in the safe agricultural production is not enough.

6.2.2 The Development and Trend of Nutritional Foods

Nutritional foods were developed under the guidance of nutrition science, which is different from the conventional food and with distinctive nutritional characteristics. Nutritional food includes nutrient-rich food and nutritionally balanced food.

The present research and application on the nutritional foods mainly focus on increasing the content of essential amino acids (lysine, tryptophan), vitamins (A, E), trace elements (iron, calcium, zinc, selenium, etc.), antioxidants (polyphenols, flavonoids, carotenoids, anthocyanidin), unsaturated fatty acids (ω-3) and so on. Foods rich in certain nutrients (such as, fiber, protein and selenium, etc.) were produced by means of agricultural production technology (such as, plants and animals "biological enhancement" breeding, fertilization, irrigation and feeding technology). Among these nutritious foods, some of them are rich in certain nutrients, such as soy protein, milk calcium and oat dietary fiber, high lysine corn, high VA sweet potato and soybean, and milk rich in selenium and so on. The production of nutritional food can upgrade the agricultural products and meet people's need for nutritional food.

6.2.3 The Development and Trend of Multi-functional Food

As the development of genomics and proteomics, some ingredients with health-care function in plants and some functional peptides in animal and human body have been gradually clarified. For instance, the ferritins distributed in rice and soy which can prevent and treat with iron deficiency anemia; soybean glycinin has lower blood pressure and lipid functions; ω-3 series of α-linolenic acid contained in sesame and perilla have functions to fall blood pressure and blood lipid, improve blood vessel elasticity, and prevent the occurrence of coronary heart disease; the vitamin E contained in sesame and sunflower can prevent coronary heat disease, atherosclerosis, cerebral softening, cancer; resveratrol contained in rape and grape can fall blood lipid and prevent thrombosis. At the same time, some proteins or peptides in animal and human body with promotional health-care function have also been found. For instance, the short peptide found in yolk can low blood pressure, the GLP-1 distributed in human body can stimulate insulin secretion and prevent and treat diabetes.

Although the agricultural products contain health-care functional ingredients, the content of these ingredients is often too low to work effectively. Using modern biotechnology to develop new varieties of plant and animal with nutrition function, and bioreactor technology to produce health functional products have become the inevitable way for the development of multi-functional foods. For example, the "Golden Rice" rich in vitamin AD, maize with high-lysine corn, and rice with high-tryptophan, iron-rich, and pig with high ω-3 polyunsaturated fatty acid content. Meanwhile, rice variety with the function of prevention and treatment of hypertension, diabetes and allergies has also been developed.

Functional foods present a new trend in the development of modern food and it is one of the important areas for the agricultural development in the 21st century. In the next 10 years, the development of functional food industry will by the yearly all exceed 10% rate. At present, the production and consumption of functional food in our country is still relatively low, and the majority of functional foods in the market are developed based on the traditional Chinese tonic, diet, or the experience and formulations of traditional medicine. Most of them are still natural primary products, and the boundaries between some of these products and traditional medicine are not very clear. The majority of health food products on the market now only have a single or mainly in one aspect of health effect, such as regulation of immune system, anti-aging, diet, losing weight and inhibition of tumor and so on. However, the effect of these functional foods is always slowly and not very obvious and the "multi-functional health foods" with two or more health effects are very rare. In order to make the research and development of our country's functional food serialization, rationalization and stabilization, firstly, we should fully explore the new functional ingredients in plants, and discuss their enrichment theory and regulation mechanisms to develop products with different functions according to the status and characteristics of functional food development in our country; secondly, we should strengthen interdisciplinary research and establish a series of functional evaluation system to evaluate the existing functional food and develop new products and new types of functional foods. We should strengthen the research of new functional food, develop appropriate regulations or management, and establish strict functional evaluation and examination system and approval system based on the international research results.

6.2.4 The Development and Trend of Food Safety Technology System

1. The technology of the quality of agricultural products and the control of safe process

The international experience shows that the realization of the whole process of management "from farm to fork" and the establishment of monitoring system from the source to final consumption are very important for food safety. The application of some advanced food safety control technology in food, such as "good agricultural practice (GAP)", "Good Manufacturing Practice (GMP)", "good hygiene practices (GHP)" and "harm analysis critical control point (HACCP)" are very effective to improve the quality of food business and the safety and quality of products. Strengthen the analysis of key factors that affect the agricultural quality during the production process, and enhance the research of pollution ways of toxic and hazardous substances in agricultural products will provide technical support for the base of pollution-free agricultural products, green food product and organic food. For the implementation of source control, some pollution ways and laws data are not sufficient and the basic research is still needed.

2. The key on-site detection technology

The rapid detection and monitoring control techniques for the residues of pesticides, veterinary drugs, food and feed additives and persistent toxic pollutants, environmental hormones and biological toxins can not meet the needs of food safety control, still needed to be improved. At present, we still lack the on-site detection technology which is urgently needed for market oversight and suitable for our production characteristics (sensitivity, fast).

3. Risk assessment technology

Risk assessment is the necessary technical means and important standards, which was emphasized by the WTO and the Codex Alimentarius Commission (CAS) used to establish and evaluate the effectiveness food safety measures. One of the reasons for our country's food safety technical measurements inconsistent with international standards lies in the wide application of risk assessment techniques, in particularly lack of the exposure assessment and quantitative risk assessment for the chemical and biotic harm. February 28, 2009, the Chapter II of The People's Republic of China food safety law precisely prescribed the content of "food safety risk monitoring and assessment". We must establish a set of new methods to assess and reduce outbreak of foodborne diseases, while strengthening the evaluation of the food-related chemical, microbiological and related risk factors, we need gradually to establish the scientific and predictable safety evaluation system that suitable for our country and continually improve it in practice.

6.2.5 The Development and Trend in Environmental Quality and Safety

United States, Britain and Germany and other developed countries carried out the research earlier on the filed of environmental quality and were in the lead of the world in the process of environmental pollution, risk assessment and environmental contamination and remediation. Those counties are strict with the ecological environment quality in agricultural production, and implementing environmental quality standards of production area rigidly. But for China, there are more people and less land; its economy is still growing rapidly and the establishment and implementation of some standards and polices still needed to be further standardized and improved.

Attention on China's ecological environmental problems has been drawn since 1970s. Domestic academia carried out the research on environmental background values and the investigation of environmental capacity, environmental risk assessment of polluted soil and water quality, environmental restoration, water quality and soil environmental quality standards and so on. In terms of soil environmental background and environmental capacity, 42 types of soil and more than 60 parameters of the basic data are included in the analysis of the environmental effects and maximum load capacity of many heavy metals and pesticides for different types of soil, a variety of plant and microorganism; carried out the basic theory and applied research on contaminated soil remediation.

The research on the regional environmental quality of contaminated soil has achieved initial success, especially in some highly polluted areas, such as sewage irrigation and areas around mining and metal company. Based on the soil environmental quality standards and risk assessment, the contaminated soil evaluation method has been set up and the classification of soil pollution has been conducted.

The research on the polluted soil control has been developed for more than 30 years. The method initially used mainly on chemical control, such as using montmorillonite and other natural or synthetic mineral, industrial slag and organic fertilizer, can significantly reduce the pollutants moving from the soil to the aerial parts of crops. However, these methods only can minimize the bioavailability of pollutants in soil, pollutants still exists in the soil and they can re-release when the environment changes. Therefore, the bioremediation technology has been valued and promoted since the late 90's last century. Many progresses have been made in hyperaccumulator screening, absorption mechanism and crop post-disposal, and the use of plant and microbial remediation technology to treat organic pollutants in contaminated soil also yielded some results.

Air pollution research in our country also has made remarkable progress. The analysis of air pollution sources, the mechanism between pollutants and surface of atmospheric particles, the atmospheric circulation of pollutants and pollution control have been carried out. The study shows that there are many atmospheric pollutants, which have exerted remarkable influence on agricultural production and environmental safety. Although DDT and BHC have been banned in China for more than 20 years, these pollutants still can be found in the atmosphere of Yangtze River Delta region.

Water pollution is seriously harmful to the safe production of agricultural products, it is necessary to control the content of pollutants in water strictly, and establish scientific and reasonable environmental standards for water quality. China has established more than 300 sewage treatment plants, 70% of urban sewage is treated.

6.3 Technology Development Goal

6.3.1 The Overall Goal

Strengthen the researches on agricultural production safety, prevention and control of major pets and diseases; maintain the nutrition of agricultural products, clean control, storage and processing and so on, the key technological breakthroughs and integrated technologies integrative innovation, the establishment of the prevention and control warning system of agricultural pests and diseases, intelligent expert management system, and form the food safety digital tracing system from farm to fork, implementation of accurate monitoring and prevention and control of "active security strategy". On this basis ① to establish a standardized system of safe production of agricultural products and green environment, to ensure the quality and safety of agricultural products in the cultivation, breeding, storage and processing; ② Comprehensive analysis of the theory of agricultural nutrients, using a fast intelligent design platform to achieve precise design and quality regulation of food; ③ create "intelligent personalized nutritional food" to meet the personalized nutritional needs according to the physiological characteristics and health status of different groups, provide the functional foods for the purpose of effectively prevent and reduce the occurrence of disease and enhance the physical quality and fitness of the whole population.

6.3.2 Phase Goals and Priority Areas

1. The development goals and priority areas by 2020

Strengthen the basic research on the related scientific issues according to the key factors that affect the quality and safety of agricultural products; with the application of breakthroughs in key technologies and the integrated application of comprehensive technologies to eliminate the risk factors that affect the agricultural products safety, establish the standardized technical system of agricultural safety, storage and processing, construct the ecological environment of green food production and provide green, safe and high quality food.

(1) The key area of the research and development

Study the mechanism of agricultural diseases and interaction between the pathogenic germ and the host; integrate the comprehensive and efficient disease control technology; study the mechanism of livestock and aquatic diseases and animal's immune defense, integrate the safe and effective disease control technology, warning and monitoring, quarantine and diagnosis and immunization techniques; study the cycling metabolic mechanism of agricultural nutrient postpartum, the mechanisms of disease occurrence and quality control, integrate a series of new storage and refreshing techniques and process and quality control techniques; study the generation and regulation mechanism of pathogenic toxins and antibiotics, form the rapid detection methods and effective control technology; study the circulating transformation mechanism of harmful substances from agricultural production area, develop the biodegradable technology of soil harmful substances, the remediation technology of contaminated soil, and the monitoring and evaluation methods of environmental quality.

(2) The key area of new product development

New environment-friendly bio-pesticides and highly effective biological agents for the control of agricultural diseases; the specific vaccine for the prevention of aquatic and animal diseases; the new green preservatives for fruit and vegetable; the additives of processed food and feed.

(3) The establishment of standard system of key area

The agricultural pest and disease early warning monitoring network and disease prevention and control system; the agricultural technology system of production safety standards; the control technology system of agricultural products storage and procession, create efficient, clean and safe management system, improve the related standards of safe production of agricultural products; the environmental background information system of agricultural, and rational environmental quality standards and risk assessment methods; establish the purification technology system of major disease areas for the main producing areas of animal food, and a platform of separation, screening and research for new green veterinary drugs and feed additives; build the techniques that with fingerprint structure information for the origin and identification of adulterated agricultural product quality; For the agricultural production process, carry out the systematic study on rapid portable monitoring technology for the risk factors of key step; establish the risk assessment model of risk factors, diet and food safety.

2. The developmental goals and priority areas by 2030

Reveal the regulatory mechanism of factors that affect agricultural products safety, promote the standardization and safe production mode of agricultural products, and realize the precise control of storage, logistics and processing quality of agricultural products; clarify the principles of restoration of contaminated environment, establish the management system of intelligent environmental monitoring and remediation of contaminated soil, create a sustainable ecological environment; study the theory of agricultural nutrients, use the accurate and rapid intelligent design platform to develop processed foods with various nutrition and provide a variety of nutrients that human health needed.

(1) The key area of information technology system

Reveal the metabolism and regulatory mechanism of nutrients of agricultural products, analyze the theoretical basis of the agricultural products, establish the precise and rapid design platform and production technology system; reveal the cycle and transformation mechanism of the environmental harmful substances of producing area, establish the technology system for the continuous, remote monitoring of environmental quality and remediation of contaminated soil, fully implement the comprehensive remediation of contaminated soil and construct the ecological environment for sustainable development; reveal the metabolic mechanism of biological toxins and the antagonistic mechanism between the insect pests and their natural enemies, improve the dynamic model of agricultural pests and diseases of early warning, establish the entire process of digital file management procedures for the agricultural production and agricultural safety, the tracking system can be traced back (TRACEFISH) and risk assessment systems, as well as the digital monitoring and early warning systems of the foods from farm to fork; clarify the antagonistic mechanism and co-evolution law of the pests and their natural enemies, establish the industrialized breeding technology system of the natural enemies; improve the early warning and prevention system of animal disease; improve the standards of application of veterinary medicine, feed and feed additives, and the monitoring system of animal food safety; establish the agricultural products on-line in situ detection technology, improve the monitoring of the key factor of production processes; establish the rapid or non-destructive testing technology of quality of agricultural products and nutrition index, according to agricultural products their own electrical and optical properties and composition changes; establish the expression database of the risk of human health, which is caused by the law of metabolism of agricultural risk factors, with the application of modern molecular biology and information science and technology.

(2) The key research and development area of diversity of nutritional food

Using high-tech bio-technology and new process technology to develop various food nutrition fortifier and multi-functional processed foods, such as rich in amino acids, protein, vitamins, mineral nutrients, trace elements (iron, calcium, zinc, selenium, etc.) and antioxidants (poly)phenols, flavonoids, carotenoids, flower pigments), to improve the diet constitution and meet the requirements of different groups for various health nutrients.

3. The developmental goals and preferential areas by 2050

Clarify the metabolic and synthetic way of various nutritional components, develop a harmonious agricultural environment which can satisfy multi-needs, lay the theoretical foundation of quality regulation of processed foods, set up the design standards of personalized nutritional foods, create "intelligent personalized nutritional food" based on the physiological and health characteristic of different groups, meet the individual nutritional needs, improve the physique and health level of all the people and provide individualized functional foods to effectively prevent and reduce the diseases.

(1) Establishment of key areas of intelligent technology system

Achieve the standardization system of techniques of accurate agricultural products cultivation, and the intelligent management system of matching technology for the treatment of commercialization, establish a complete digital food safety standard, testing and network monitoring system, and comprehensively enhance the quality of agricultural products and the market value of goods; expand the monitoring area and density of agricultural environment, comprehensively restore the contaminated production environment, and restore the quality and function of the contaminated environment; establish the intelligent network control system to monitor and prevent the pests and disease of agricultural products, develop new animal and plant specific resistance inducer, green biological agents and establish the standardized system of industrial production techniques to create accurate and efficient techniques target to the natural enemies and pesticide, and the intelligent system of pest early warning, prevention and control technology; establish the standardized green cultivation technology system of animal production, and the quality control and monitoring system throughout the animal production. Develop the human health monitoring and rapid diagnostic technology, intelligent assessment and design the formula of nutritional requirements; establish the high-throughput, highly sensitive detective technology of functional nutrition according to the properties of agricultural products and conditions of production areas.

(2) Key areas of development of personalized multi-functional food

Use accurate, fast and intelligent design platform, according to the physiological characteristics of different groups (such as the elderly, infants, students, and women), to intelligently design the personalized functional foods, develop "intelligent personality nutritious food"; according to the different human health status (such as hypertension, hyperlipidemia, diabetes, reduced immunity, atherosclerosis, hyperthyroidism and senile dementia, etc.) to develop "personalized functional food" and the related processing technology, quality intelligent monitoring system and intelligent production management system, meet the individual nutritional needs, provide various and personalized functional foods for the effective prevention and reduction of disease and improvement of the health of all citizens. Establish the functional assessment and safety evaluation system of functional food through the research on its metabolism and mechanism of function and process monitoring.

6.4 Technology Development Roadmap

6.4.1 Science Mission

In view of poignant problems in the food safety of our country, aiming at the research focus and development trend in this field, combined with our country's science and technology base, the scientific missions of future food safety technology include: the metabolism and regulation mechanism of agricultural nutrient; the interaction mechanism between the defense and pathogens of animal and plant and their host; the metabolism and regulation mechanism of biological toxin; the antagonistic mechanism and co-evolution law of pests and natural enemies; the cycling and metabolic mechanism of agricultural products nutrient; the cycle and transformation mechanism of the harmful substances in the production areas.

At the same time, the development of the following nine core technology and system need to be supported: the technical system of standardized safety production of agricultural products; animal and plant stress resistance inducers, the new vaccine and drug development technology; the scale rearing technology of natural enemies of plant pests; the research and development technology of biological agents and bio-pesticide, the control and risk assessment techniques of contaminated environment; the development technology of multi-nutrient food; the intelligent design technology of personalized functional foods.

6.4.2 The Design Thought of Roadmap

Around the science mission mentioned above, fully absorb and use the computer network technology, 3S technology, computer visualization, geographic information system, high-quality satellite images, BP neural network, intelligent expert systems, high-precision technology management, establish the early warning and monitoring system, and intelligent expert management system of animal pests and diseases, and set up the prevention-based comprehensive prevention and control of food safety technology and management system; establish the food safety digital tracking and warning systems from farm to fork, carry out the "active food security strategy" with accurate monitoring, pre-emptive "disease" and precise prevention and control.

Based on the comprehensive analysis of all elements of animal and plant security products, with the use of the precision and rapid high-throughput based detection and intelligent design platform, develop "intelligent personalized nutritional food", to satisfy all the people's permanent need for food security and nutrition.

6.4.3 The Overall Roadmap

The roadmap and sequence graph were drawn (Fig. 6.1, Fig. 6.2) by the integration of various elements, such as needs, tasks, core technologies and platforms, with three development steps:

Before 2020, focus on the clarification of the metabolism and regulation mechanism of agricultural nutrient and bio-toxins; accelerate the establishment of accurate and rapid high-throughput detection platform, develop the pest control technology and security storage and processing technology; establish and optimize the standardized technology system of safe production; eliminate the hazards that affect food safety, form the ecological environment of agricultural products. Set up the tracking system (TRACEFSH) that could be traced back and risk assessment systems, as well as the food safety digital tracing detecting and early warning systems of food safety from farm to fork.

Before 2030, the research of key fields focuses on the diversity demand of nutritious food. The establishment of the early warning dynamics model and monitoring network system of animal and pests diseases, intelligent expert systems and the theoretical system of the animal and plant system acquired immune mechanism, start the 3S-based security monitoring and early warning and intelligence experts and production management platform. Develop the animal and plant stress resistance inducers vaccines; develop the target release technology of chemical pesticide of precise and efficient predators, biological pesticides, and low toxicity through the study of natural enemies breeding technology, high specificity and low toxicity bio-pesticides, bio-antagonist and chemical pesticides with high efficiency and low toxicity. At the same time, develop the multi-functional animal and plant food and nutritious food. Based on the basic research on the immune function in breeding animals, the regulation in the response process, the molecular mechanism of targeted drugs, and the basic research on the engineered vaccine, animal waste heavy metals and the ecological toxicology of pathogenic variability to develop the new technologies and promote the generation of new technology growth.

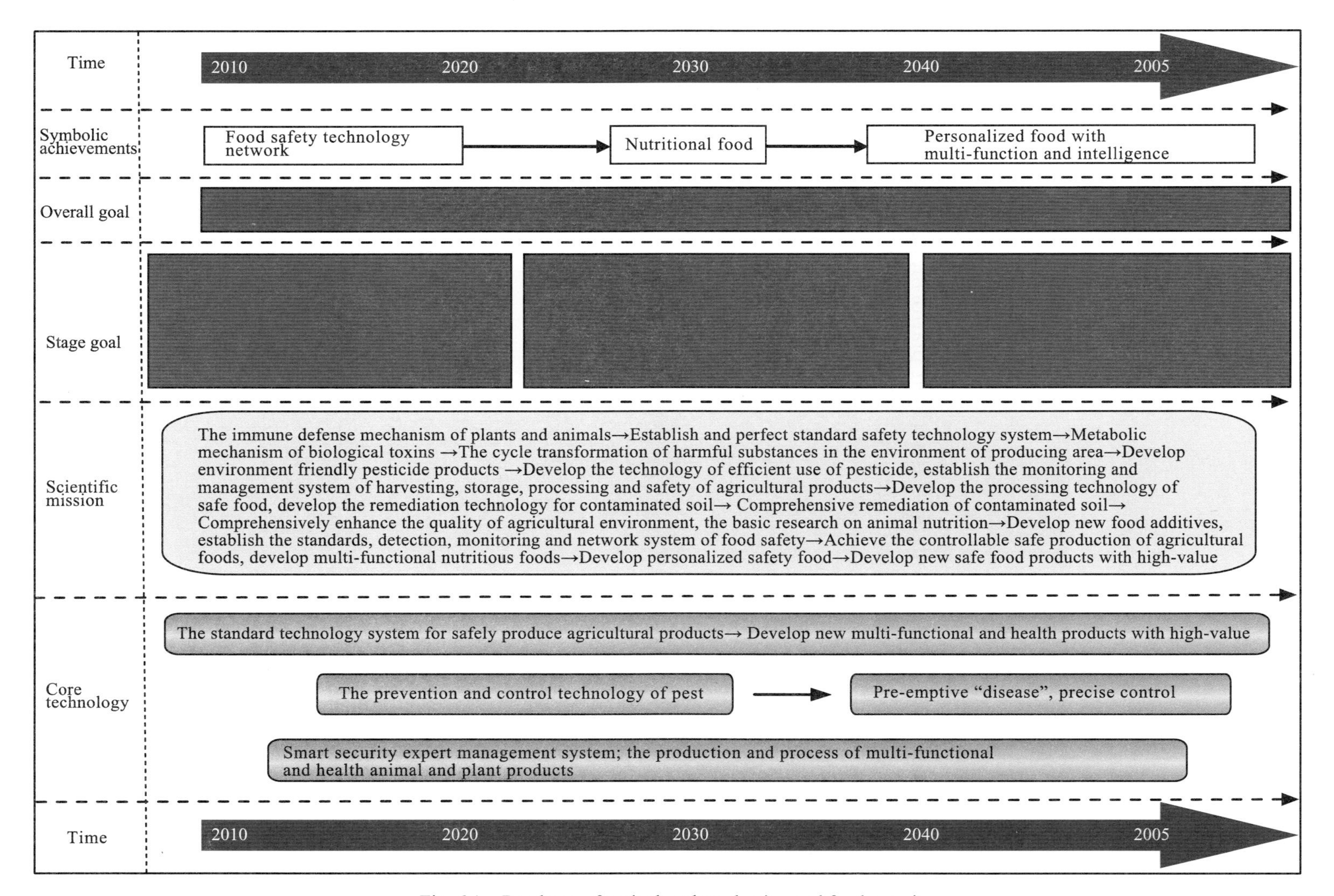

Fig. 6.1 Roadmap of agricultural production and food security

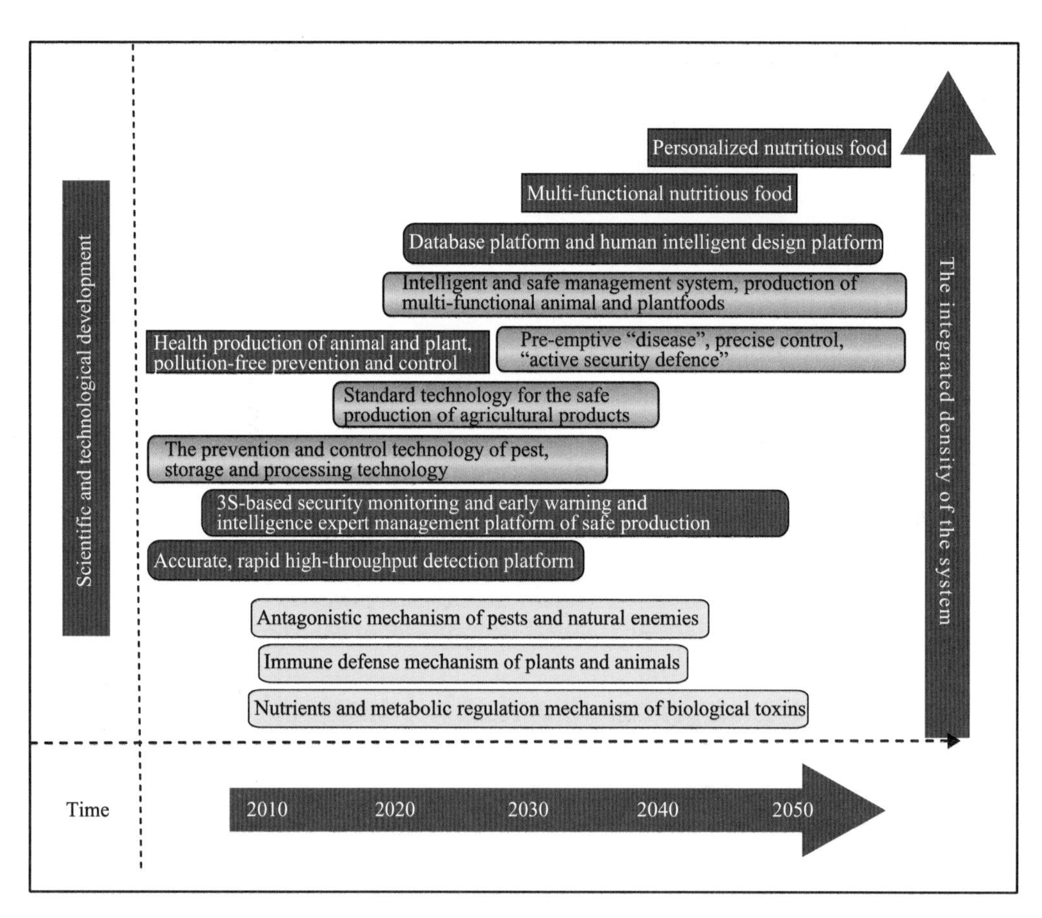

Fig. 6.2 Timing diagram of agricultural production and food security

Before 2050, the key field is to meet the individual needs of multi-functional food. To make major progress in some important scientific issues, such as the interaction mechanism between the animal and plant immune defense, pathogens and the host, the antagonistic mechanism and the co-evolution law of pests and natural enemies. Realize the strategic shift in the field of food safety from the "passive control" to "active safety defense". Strengthen the dynamic early warning model and the monitoring network systems of animal and plant pests, conduct the monitoring, early warning, diagnosis and prevention and management of the trends and outbreaks of pest and disease. Through the intelligent data platform and human intelligent design to develop the "personality nutritious food", and establish the "active security defense" system of animal and plant pests and diseases, through the development of natural enemies, resistance, biological control agents and high-precision insect sex pheromone agent to achieve the goals of pre-emptive "disease" and precise control.

6.4.4 The Key Technical Programs to Realize the Roadmap

1. The common technologies of food safety programs

Establish the early warning monitoring and intelligent expert management system of pests based on the 3S technology, develop the new technologies of prevention and control of pests, develop the environment-friendly production technologies of biological agents; establish the new technology for the "preemptive diseases, precise treatment" of pests, ① adopt the early warning dynamic model and monitoring network system to monitor, early warn, diagnosis and prevent the epidemics and outbreaks of pests and diseases; ② improve their own ability of animal and plant to withstand adversity and immunity, universal application of animal and plant stress resistance inducer and animal and plant vaccines; adopt the integrated pest control techniques, such as the precise release of natural enemies, veterinary drugs, biological pesticides, sex pheromone and low-toxicity chemical pesticides, to achieve comprehensive and lasting security of plant and animal production.

Develop the new storage and preservation technology of animal and plant products, the precise and rapid high-throughput testing technology of quality and safety, the bioremediation technology of environment, the technology with safety and high quality factors optimally allocated, set up the intelligent technology systems of the production of animals and plants products, and the products treatment technology system.

Strengthen the new technologies to design and develop intelligent personalized nutritious food, produce new types of multi-functional and personalized nutritious food, improve the people's diet structure and health level.

Create the control technology for the major pest and disease and environmental safety, the metabolism regulation of animal and plant nutrition, and the innovative research platform of database and human intelligent design. Upgrade the traditional cultivation and breeding techniques comprehensively, form the innovative research and industry groups with obvious advantages and outstanding features in the field of animal and plant production technology. Lead the transformation and upgrading of technology systems of animal and plant products in our country, provide the proactive technical support for the establishment of efficient, safe and modern system of animal and plant products. Comprehensively enhance the overall quality and developmental capacity of the technology of animal and plant products in our country, make the scientific research and production technology of our country in this filed reach the world advanced level.

2. The safety program of plant production technology

Establish early warning dynamic database of plant pests, develop intelligent expert pest management system. Develop plant stress inducers, plant-based vaccines and other biological control agents. Develop the advanced synthetic technology of high-precision pest sex pheromone; develop the efficient and controllable biological control technology; develop the application technology and its integrated supported system for prevention and control technology with high specificity and low toxicity biological pesticides, chemical pesticides, biological antagonist and accurate target technology. Make sure the residues of toxic substances, drug and heavy metals in the plant products to be minimized. Develop the safe food additives and preservatives; establish a complete food processing safety standard, testing and monitoring network system.

3. The safety program of livestock production technology

With the application of nutrition, cytology, histology, molecular biology, botany, ecology and even the advanced technology of information science, the advanced technologies of immunology, genomics, genetic engineering, virology, microbiology and bioinformatics will be applied in the technology area of major animal disease prevention and control. The advanced technologies of microbiology, environmental chemistry, ecotoxicology, protein engineering, enzyme engineering, fermentation engineering and systems engineering will be applied in area of environmental safety technology, through the interdisciplinary and technology combination to establish the safety innovation technology platforms (such as the National Key Laboratory) of modern livestock production and animal products, Achieving great discoveries in the basic theory of security studies on the modern livestock production and animal products, has significant breakthroughs in research methods and major innovations in technical aspects of regulation. Adopt the explore means of new genetic resources to develop the core technologies of planting grass and natural pasture, break through the bottleneck that hinder the development of animal husbandry in northern China, rapidly enhance our country's innovative research capability of grass.

4. The safe technology program of aquaculture production

With the use of biotechnology and information technology to conduct long-term, continuous dynamic monitoring, repair and evaluation in the operations of precise fishing, aquaculture resources, ecological systems and environment. Explore and utilize resistance of living aquatic resources to achieve accurate high resistance breeding; develop the processing and feeding technology of pollution-free and efficient feed, research and develop the automatic feed metering technology and technology of feeding on specific time. Establish the technology of rapid detection of pathogens and the diagnosis and early warning and forecast technology of disease of aquaculture. Screening the immune stimulants and new harmless fishery drugs; develop the vaccines for the virus and bacterial disease.

5. The safety technology program of production environment

The ecological environment safety is the foundation and guarantee for the agricultural production safety and food safety, only the safety of environmental quality of agricultural origin can conduct safe agricultural production, avoid producing environmental pollutants from the soil or water to the biology and enlarged in food chain, which can cause the agricultural pollution and damage to human health. Develop the pollution monitoring, control and remediation technology of agricultural ecosystems, combined with information technology development, put forward our country's agricultural safety policy and management practices, fully establish the safety of agricultural systems.

(1) The quality and safety of soil environment

Carry out the investigation of the soil environmental quality and the study of pollution process of the origin of agricultural products; conduct the assessment of soil environmental quality; raise the standard procedures and methods for the investigation of soil background, establish the informational system of soil background, and discuss the relationship between the quality and safety of agricultural products and soil background environment. Establish the methods of utilization for different soil, the scientific and reasonable soil environmental quality standards and risk assessment methods; develop the agricultural safety planning and zoning; construct the monitoring technology and system of soil environmental quality and its early warning system; the technology based on the fixation and blocking of soil contaminants, and the methods of microbial collaboration and degradation, research the techniques and principles of pollution control and restoration to reduce pollutants transfer from soil to plants. Effectively control the pollution sources, protect the environmental quality of agricultural origin, and restore the contaminated soil in our country gradually, meet the security of crop and animal production.

(2) The water quality and safety

For the key areas or key territorial water, to carry out the investigation of water quality and the research on the pollution process, conduct the assessment of water quality and zoning. Establish the scientific and reasonable standards of water quality and risk assessment methods; set up the planning and zoning of agricultural products production; construct quality monitoring technology and system; strict implementation of the wastewater discharge standards, develop the related technologies of water pollution control, and gradually improve the water quality of rivers and lakes in our country, meet the safe water supply of crop, animal and aquaculture.

(3) The safety of air quality

Carry out the investigation of atmospheric environmental quality and research on the pollution process, to analysis and track the pollution source and evaluate the atmospheric environmental quality. Establish the scientific and reasonable standards of atmospheric environmental quality and risk assessment methods; develop the agricultural safety planning and zoning; Construct the regional air monitoring technology and system networks; control the pollution source strictly, develop the related control equipment and technology to reduce the atmospheric environmental pollution, improve the quality of atmospheric environmental conditions step by step, meet the safety of the crop, animal and aquaculture.

Main References

[1] National Research Center for Environmental Analysis and Measurements. The memorabilia of dioxin in China. The research on dioxin, [2004-09-30]. http://www.cneac.com/article/list.asp?id=63

[2] Europe's major food safety incidents in recent years, Journal of Chinese Institute of Food Science and Technology, 2008, 8 (6): 47

[3] Wang Y. The overview of infection and pandemicity about Escherichia coli O157:H7. Progress in Microbiology and Immunology, 2008, 36 (1): 51-58.

[4] WHO Media centre, Food safety and foodborne illness, Fact sheet, [2007-03]. http://www.who.int/mediacentre/factsheets/fs237/en/.

[5] Daszak P, Cunningham AA,Hyatt Al. Emerging infectious diseases of wildlife threats to biodiversity and human health. Science, 2000, 287: 443-449.

[6] Binder S. Emerging infectious diseases: public health issues for the 21st century. Science, 1999, 284: 1311-1313.

[7] Pearson H. SARS: what have we learned. Nature, 2003,424: 121-126.

[8] Guan Y. Isolation and characterization of viruses related to the SARS coronavirus from animals in southern China. Science, 2003, 302: 276-278.

[9] Capua H. Alexander DJ. Avian infuenza and human health. Acta Tropica, 2002, 83: 1-6.

[10] Nichol ST. Ariwaka J, Kawaoka Y. Emerging viral diseases. PNAS, 2000,97: 12411-12412.

[11] Ferguson NM, Fraser C. Donnelly CA, et al. Public health risk from the Avian H5N1 influenza epidemic. Science, 2004, 304: 968-969.

[12] Hinshaw VS, Webster RG, Naeve CW, et al. Altered tissue tropism of human-avian reassortant influenza viruses. Virology ,1983, 128: 260–263.

[13] Kilpatrick AM, Chmura AA, Gibbons DW, et al. Predicting the global spread of H5N1 avian influenza. PNAS ,2006,103(51): 19368-19373.

[14] Li KS, Guan Y, Wang J, et al. Genesis of a highly pathogenic and potentially pandemic H5N1 influenza virus in eastern Asia. Nature, 2004, 430: 209-213.

[15] Wang C, Feng Y, Pan X, et al. SalK/SalR, a two-component signal transduction system, is essential for full virulence of highly invasive Streptococcus suis serotype 2. PLoS ONE, 2008, 73(5): e2080.

[16] Feng Y, Zheng F, Pan X, et al. Existence and characterization of allelic variants of Sao, a newly identified surface protein from Streptococcus suis. FEMS Microbiol Lett, 2007, 275(1):80-88.

[17] Chen C, Tang J, Dong W, et al. A glimpse of streptococcal toxic shock syndrome from comparative genomics of S. suis 2 Chinese isolates. PLoS ONE, 2007, 2(3): e315.

[18] Tang JQ, Wang CJ, Feng YJ, et al. Streptococcal toxic shock syndrome caused by Streptococcus suis serotype 2. PLoS Medicine,2006, 3(5):668-676.

7 Roadmap of Agricultural Modernization and Intelligentization Science and Technology Development

Wide application of Information Technology (IT) will reconstruct the framework of world socio-economic development, as all sectors, to 2050, will be informationized. As for agricultural sector, wider application of IT seems to be more difficult than other sectors, because of its features as highly dependent on natural resources and related to the production of living materials, with high temporal and spatial variability and low controllability and stability. Nevertheless, such extraordinarily complicated system, to the point, urgently needs the support of IT to realize agricultural intelligentization. Only with the development of digital, precision and intelligentized agriculture mainly characterized by information elements, the traditional production patterns can be broken through and precise management of agricultural production can be realized to largely enhance the efficiency of agricultural production and resources utilization and the yield and quality of agricultural products, to save costs, to protect environment, and to speed up the circulation and trade of agricultural products, and finally to reach the level of agricultural modernization and intelligentization.

7.1 Development Needs, Significance and Tendency

7.1.1 Development Needs and Significance

Agriculture, served by IT and equipped with intelligentized technology, cities and countryside, connected by network technology, and the new countryside, digitalized by digital technology, are necessary options for the future agricultural development in China. The development of agricultural IT and precision agriculture is not only conducive to the promotion of our high-tech agriculture with self-Intellectual Property Rights, but also of strategic significance for the advancement of our agricultural modernization level and the improvement of our overall agricultural competition ability in the world arena. To 2050, techniques as quick acquirement of agricultural information, data transportation, network for wireless sensors, electro-labeling, surveying, agro-expert decision system, artificial system, and intelligent instruments will become the core requirements in agricultural informationization and precision agriculture, thoroughly renovating all courses of the agricultural processes.

1. Development of modern agriculture requires the support from network-based, standardized and synchronized agricultural information service system

Fluent data transfer in various levels and scales is an important token for agricultural modernization. Through several years' efforts, information expressway has been primarily founded, mainly based on the Internet and assisted with networks of broadcast, telecom and satellites, with a total data resources and websites related to agriculture of 1,000T and 10,000, respectively. However, on the way to agricultural informationization, we are still confronted with lots of issues to be addressed, lagging techniques as data collecting, processing, and analyzing lead to slow information update and release; dispersed data resources with a low standardization level results in great difficulty in effective resources integration and low use efficiency; low interaction of different information networks blocks the channels for information service, and farmers can not acquire whole information easily; the products of agricultural information service were not farmer-friendly and limited its wide use. Therefore, network-based, standardized and synchronized agricultural information service system is the future technological requirements. Such system should be integrated with multiple networks such as the Internet, telecom, satellites, etc.; standardized, multi-media and shared data collecting, processing, storing, accumulating and serving of agricultural information resources, especially the basic agricultural information; portable, diversified and low-cost service terminals; and with the new techniques as intelligent searching, gridding, IPv6 integrated and applied.

2. Precision and intelligence technologies as required by high-efficiency modern agricultural production

In China, arable land and water resources are in great shortage, and the current unreasonable fertilization and irrigation practices and excess pesticide application sustaining the high yield not only induced a great amount of waste, but, importantly, polluted the agro-ecosystem and affected the quality security of agricultural products. Thus, with the target of increase in yield and quality and improvement in environment, the development of "reasonable plantation and cultivation, resources-saving, environment-friendly" precision technologies, linked by agricultural information technology and integrated with modern plantation and management strategies, will update the agricultural production patterns, transform the traditional extensive agriculture into the precise one, reduce the resources consumption, environmental pollution and production costs, and elevate the output and production rate. Moreover, agricultural development provides a platform for intensive application of various new technologies, materials and methods, and its scale and intensity will largely enhanced, and thus the agricultural intelligentized equipments will be greatly in need. The development of versatile, cost-effective and intelligentized facilities meeting the requirements of modern agriculture, such as auto-navigated tractors, variable tumbrels, intelligentized pesticide sprinklers, high-efficiency seeding-machines, sprout-boosting instruments, agricultural robots, etc., is of great significance to modern agriculture.

3. Digital technology provides effective tools for monitoring and warning natural disasters in the agricultural sector

Agriculture is a resources-based industry, so the key to realize sustainable agricultural development lies in the balance between resources consumption and protection. Over-consumption of resources will lead to unsustainable agricultural development, and over-protection will result in stagnant agricultural development. Therefore, on the one hand, a good master of agricultural resources is the pre-condition for sustainable agricultural development with the application of IT, on the other hand, agricultural disasters in China are the most severe and regular among the world, especially the typhoon, hailstone, ice snow, sandstorm, heat wave, drought, floods, plant disasters and insect pests, etc., which have made tremendous losses to the agricultural production. The monitoring, forecasting and warning systems aimed at fatal natural disasters have become to the key projects in resisting natural disasters and maintaining agricultural production, and geographic information system (GIS), remote sensing (RS), global position system (GPS), and simulation technology would provide effective tools for such system construction. Great progresses have made in digital theories and technologies as agricultural resources management and disaster relief, but related information was sectioned among different divisions. So shared information and services in agricultural resources and disasters are urgently required to address the overlapped construction and information isolation in resources management and disaster prevention. Furthermore, In our country, agricultural resources management is separated from use, agricultural disaster prediction is separated from prevention, which needs corresponding professional model to break the separation. Therefore, studying relative technical theory of all kinds of agricultural resources, disaster information monitoring and early warning system, efficient management strategy etc, structuring digital service platform of agricultural resources management and disaster defense, realizing digital management by using digital technology have important theory and practical meaning for agricultural security and country development.

4. The development of agricultural Hi-Tech industry depends on the propulsion of agricultural information technology

The agricultural Hi-Tech industry has emerged with the development of agricultural information technology. The exploitation of application software in agricultural field, such as management of farms, crop cultivation, feed production, enterprises management engaged in processing agricultural products, farmland water conservancy and management of forest etc., improve the efficiency of agriculture and the quality of products, and profit the farmers. Agricultural Software is one of the important constituent parts of software industry. In international, especially in developed countries, agricultural software is known as "the agricultural software industry", and has already exerted increasingly important role in agricultural production.

Developing the equipment of the modern digitization technology to collect and implement agricultural information, automatically control agriculture by method of integrating information technology, engineering and manufacture technology is a necessary approach for utilizing modernization of agriculture, such as all kinds of sensors, digitizing agricultural mechanical equipment and environment control system et al. The trend of industrialization exploitation is obvious that can not only promote the science and technology standard of agriculture, but also encourage the development of relevant manufacturing industry and information service industry.

5. The requirement to structure the virtualization research platform for massive agricultural study activity and complex design optimization of agricultural productive process

Virtualization study is a network connection among cooperation partners of scientific research by applying information technology, form a informational, open cooperated scientific research environment to make a mass of manpower, instrument and information regularly connect and form a powerful research system by using advanced network facilities and information mediums for massive agricultural study activity and complex agricultural process design optimization project.

The depth and width of future agriculture research is becoming larger and larger, the system is also becoming more and more complicated. The mutual cooperation is an important mean for scientific innovation, which needs more much scientific research institutions, laboratory and scientists distributed all over the national even worldwide to work together, using historical data, experimental data, model and method extensively, even using lots of expensive and advanced instruments and equipment, but at present the methods of communication and cooperation among agricultural scientists still rest on telephone, symposium, visit research and so on, it is difficult to satisfy the development of cooperative research requirements momentarily and all-weather working. Therefore, it will be very important to establish informatization implementation scientific research environment which oriented to the agricultural field, and provide convenient information mediums and cooperative research of virtualization platform for agricultural scientific research activities.

7.1.2 The Development of Agricultural Informatization and Precision Agriculture

The development of world agricultural information technology by half century has characteristics as following: With the base of information expressway, agricultural information network system has been established, and international communication of agricultural information has been realized. All kinds of database systems applied to agricultural technology service provide information for crop growth situation, diseases and pests prevention, control technology and agricultural production materials market. Satellite data transmission system is widely applied, provide lots of including meteorological satellite image, long and short term weather prediction and product information. The comprehensive development of expert system, model system, brainpower information system have launched the system of agricultural production and operating management, including crop simulation system, crop production management system, disease and pests management system, agricultural expert system, agricultural strategy support system etc. Remote sensing, geographical information system, and global positioning system (RS, GIS, and GPS, short for 3S system) are used widely in agriculture, especially in monitoring agriculture and precision agriculture. Information and automation technique intelligent equipment system is entering research and developmental phase.

1. Agricultural information service network

Throughout the development of developed countries and our country agricultural information technique, it is uneven among developed countries, and they have respective keystone of development and advantages. With the powerful economy and scientific and technological strength, America takes world trends in almost all aspects of agricultural information technique, the developments of agricultural information technique in every country refer to American development, America puts forward new concept and new technology which have a enormous impact in the world; 3S technique of Canada is comparatively mature relatively; The development of agriculture information technology in France starts late, however, it develop fast depended on its science and technology strength and the national quality, especially in the construction of basic facilities and the research on virtual agriculture.; Auto-control technology of computer is applied widely in German; In Japan, the connection between computer and mobile communication takes more attention; The application of management information system and consultation systems for long distance are focused in Korea; In China, the experts system developed fastest and applied most widely.

A common character of developed countries is that all those counties have constructed their complete information system, especially in American, Canada, and French. The information system of American is not only criterion and standard, but also reasonable for application, especially the perfect system would improve the work efficiency, and which also could help to control the development of the information technology. The information diversity in Canada and French make use of all kinds transmit channel, provide better information service to farmer. The computer popularization rate is more high in German and Japan, and provide better information sever to farmer. Information system of Korean is still in rapid development period, although the application of computer belongs to a lower level of management information system, it is suitable to their national conditions, and is able to improve their work efficiency.

The basic theory, key technology research, product development and application of technology of Chinese agricultural information has made an important breakthrough by many years efforts. The professional service organizations in rural areas have been appeared, which was main based on internet, and radio, networks, telecommunications networks, satellite network played a subsidiary role. However, the development of information technology system in China is slow, and the construction of basic facilities is also difficult to achieve the requirements of agricultural informatization. A number of agricultural websites have been constructed, but the information is less with low use efficiency, and some websites are constructed repeatedly. In the future, the development direction of agricultural information are: intelligent, network, and multi-network integration; the standardization, multimedia and shared of agricultural services resources; the portable, low-cost of the information service terminal of agriculture; applications of information technology, such as intelligent search, web, etc.

2. Planting management informatization

Seeding informatization is a relative mature part of planting informatization in developed countries, which characterized by: ① seed enterprises management informatization: seed industry management platform system has been established, and breeding, variety testing evaluation, seed production and marketing of the whole process have achieved scientific management, the professional equipment with the main types of seeds, ERP and OA system are integrated completely. ② Information services to seed users and retailers: According to the requirements of users and environmental conditions, a special information system has been established to provide the services of varieties or seed product selection, which significantly reduces the risk of new varieties choosing and difficulty of the new varieties application. ③ Information of planting supervision and management: according to international standards, some information technology have been realized basically on the aspect of the varieties right, new varieties, certified seed, international trade and daily administrative management of seed, etc., through on-demand regulation. There have been many studies in theory of China's growing information technology, and some innovations have been obtained (such as the breeding resource groups, heterotic patterns, the high stability coefficient of variety evaluation, the rank analysis, DTOPSIS, application of GIS), but little have been produced.

After 30 years development, crop production information management system has been popular applying in crop planting management, horticultural facilities management, plant protection, breeding and economic strategy, etc., and presented a well development prospect. Chinese crop production management information system research has achieved the international standard, and a number of applied models have been developed, such as the diagnosis system for ginseng pest based on case reasoning, control and monitoring expert system for greenhouse potato growing, the agricultural remote diagnosis platform based on knowledge, the poultry disease diagnosis system based on Bayesian network platform, intelligent management system for high-quality wheat, expert system for plant diseases and pests, fertilization expert system based on Web and GIS, intelligent strategy system for rotation of wheat and corn, weeds information platform based on component, database network system for animal genetic resources, agricultural meteorology information system, etc. In addition, a series of systems have also been developed, such as agricultural resource information management system based on ArcGIS Engine, agricultural resource information database system, agricultural resource information issue system based on Web and GIS, agricultural resource management strategy support system in county scale, distribution intelligent system for gardening plant, crop varieties dynamic intelligent system based on Web and GIS, ecological health intelligent monitoring system for greenhouse crop, computer aided design system for low pressure pipeline irrigation network, virtual scene simulation system for irrigation water distribution, etc., but the application of these systems still need further examination.

Along with the development of information technology, the agricultural production environment and biological information monitoring in nondestructive, real-time combined multiple function have made much progress. The farming model and strategy system presents the trends from local to systematization, digital, intelligent, and from experience to application.

3. Breeding informatization

The application of information technology in breeding has been popular to livestock, aquaculture, breeding and economic strategy, etc. There are some representative information technologies such as Pigchamp, Pigwin and INSIGHT,etc.

In the breeding of livestock, poultry and freshwater fish, some systems have been developed in recent years, such as computer expert system of production management, animal genetic resource database network system, however, their production data and parameters are based on the traditional extensive breeding way, and their statistical model and computer components are also not precise. At present, with gradual improvement of people's life quality, people have a higher requirement on animals (water) product safety and quality. Feeding manners have been changed from traditional breeding to fine healthy breeding mode, and the computer network information technology has also presented a new progress. Therefore, the past developed expert system become difficult to adapt to the new mode of healthy breeding. GBS is a single machine version software for pig breeding, a number of expert systems are also s single unite version, there are online version research in recent years, but they haven't accepted by the market because of lack of technological integration.

In recent years, some foreign softwares have been introduced and they have no promotion with high application price and differences with the actual situation in china. At present, healthy breeding ways are carried out through out the nationwide, the new computer management system is needed urgently to adapt to the new model, which could provide the guides of the development of healthy breeding.

The development of the information in breeding is based on the critical control point during the breeding process taken as breakthrough, then developing a number of monitoring and control equipment, developing intelligent system for different stages in breeding process and control system of the animal health breeding technology, in order to improve agricultural informatization level and agricultural production management level, finally realizing precision and intelligent in the breeding production process.

4. Digital management of agricultural resources

The development of spatial information technology, which based on geographic information system, remote sensing, global positioning system and the model, provides technical support for the digital management of agricultural resources. At present, satellite remote sensing images of low resolution, which combined with GIS, using to monitor, evaluate and map agricultural resources, such as forestry, grass, land use / land cover and soil, has become a major means of investigation and monitoring of agricultural resources. Agricultural resources management will change from focusing only on the number of management to both quantity and quality of agricultural resources, management and use of integrated management paradigm. The establishment of national, provincial, county and township four agricultural resource database, the corresponding database standards, a database of agricultural resources building and data integration technology framework model sharing, data sharing implementation strategies, breaking the agricultural resources of data fusion theory techniques, the development of agricultural resources, efficient use of application models, the development of on-demand service platform for digital management of agricultural resources, the full realization of digital management of agricultural resources, will become the main trend in the next 30 to 50 years . The development of agriculture can overcome the difficulty of "resource limitations", and it is able to make the maximum agricultural output by consuming a minimum of resource and the lowest ecological and social cost.

5. Precise agriculture technology and equipment

Internationally, it has been widely researching in precise agriculture, such as farmland information collection, analysis and strategy, and precise working technology etc. Some breakthroughs in quick collection technologies have been obtained, these technologies including collection the information of soil nutrient and moisture, crop growth and physiological parameters, distribution of insect pest, and some relative technological products have been manufactured; the crop simulation models and agricultural expert systems have been established based on strategy analysis technology of precision agriculture. Precise operation technology and equipment in developed countries have been developed maturely, various electronic monitors and control equipment have been applied in complex and intelligent machines, such as variable seeding-machine, fertilizer and pesticide machine, in addition, these machines have gradually entered the international market. The application of precise agriculture technology has realized the effective utilization of agricultural resources and improved the comprehensive benefit of agricultural production.

The research of precise agriculture in China is the stage of studying and demonstration. In recent years, we have carried out efficiency research on the key technology in precise agriculture about information acquisition, information processing, variable application, developed the corresponding products and system through the system integration, applied a range of demonstration, achieved remarkable economic, social and ecological benefits, however, the shortage overall is also existed compared with the developed countries. Compared with America, China lacks the whole precise agriculture technology system which is suitable for the country situation. Compared with Japan, China lacks independent property of core technology products, especially in the aspects of intelligent and horticulture, which can't satisfy the major requirements of the new rural construction and the development of modern agriculture.

It could be forecasted, with the life science, information science, materials science, environment science and control science developing continuously as well as it is fully applied in the field of agriculture, the precise agriculture technology and correlative modern agriculture production equipment, especially the agricultural intelligent equipment, will be developed rapidly. The precise agricultural technology product which suits our country national condition, including the agricultural robot will enter the domestic and foreign markets, and it is necessary to develop gradually the agricultural technology equipment from the traditional function to the informationization, the intellectualization, the universalization, precise and the multi-purpose directions.

6. The network platform of agricultural virtualization research

In recent years, many countries have made great progress on the virtualization research. In 2000, the British scientists put forward the e-science concept that established a brand-new informationization scientific research environment based on the internet to solve the issue scale and the extendibility in the scientific research, and to realize sharing among different node different experiment platform data and equipment. In 2004-2014 year British science plan, it is explicitly proposed to promote e-science as the scientific research informationization infrastructure. From 2005 to 2009, the American Natural sciences Foundation (NSF) will keep investing 120 million dollars to support the research, development, application, platform construction and maintenance of the network. South Korea also plans to invest 1,027 billion wons to construct new informationization scientific research environment by 2010. Chinese scholars in 1998 had made the construction hypothesized scientific research environment tentative plan. In 1999, the national "863" program had started the key research project of "national high quality calculate environment". In 2003, the NSFC started the key research program, which was "Research on science activity environment based on network". In 2008, the ministry of education had set up key technology research of virtual experiment teaching environment and key research project of application and demonstration.

At present, the construction of informationization scientific research environment has already become the foundational work mission of various disciplines domain. Some construction achievement has already been operational, such as e-science, china-OV. Therefore, in future, putting virtualization research into agricultural research and to establish the virtual agricultural research platform will be the development trend of agricultural science research.

7.2 The Goal of Scientific Development

7.2.1 Overall Goal

Through the breakthrough of the key technologies and equipped by high-tech, Chinese agriculture should realize the followed goals:

1. Agricultural information service network

We should complete the common agricultural information supporting platform for network software and agricultural information network system construction for servicing the "three rural", achieve agricultural information service network, circulation of agricultural products modernization and agriculture macro decision-making scientific, promote the full penetration and widely using of information technology in agriculture and rural areas.

2. Digital management of agricultural resources

Based on space technology, remote sensing technology, sensor technology, GPS, GIS and intelligent technology etc., we should overcome key technology of information collection, complete the construction of the digital system of agricultural resources, such as soil, water resources, climate and so on, comprehensive monitoring of the dynamics of agricultural resources and agricultural ecological environment, early warning of natural disasters such as weather and pest, estimation of crop growth area, growth and yield.

3. Precision management of the agricultural production process

We should complete the construction of information collection system, simulation models, management strategy system during the process of plant and animal production. Combined with intelligent equipment, precision agricultural management of the production process should be achieved, and which greatly improve the efficiency of resource use and production.

4. Intelligent of agricultural equipment and agricultural machine

We should complete some important intelligent equipment, precise agriculture machine and system prototype development, which can be applied in major field crops and production of facility agriculture in China, and can be used largely in agricultural production. Then, we should enhance the modernization of agricultural equipment and mechanization levels of our country.

5. Network platform for virtual research in agriculture

We should establish a network platform for virtual agriculture research with well autonomy, interactivity, scalability and security, achieve efficient resource aggregation and widely sharing of experimental instruments, data resources, computer resources and services based on internet, solve some problems, such as imbalance research resources, inconvenient cooperation, and limitations of geographical, time, weather to agricultural research activities, promote teamwork, provide strong technical support for agricultural research and technological innovation.

7.2.2 Stage Goal

To year 2020: We should complete the development of multi-functional agricultural information network platform and professional search engine, establish a large regional scale professional databases, especially resource archive database, model base and refreshable system, complete agricultural information service network, provide services of regional resources optimization, production configuration, technical consulting, market demand analysis, product traceability, etc., study key technologies of digital resource management, precise management in manufacturing process, intelligent agricultural equipment, obtain breakthrough in remote sensing of disaster monitoring, drought monitoring, crop growing monitoring, soil moisture information collection, precise seeding, precise irrigation, etc., study the prototype of the network platform for agricultural virtual research, identify the feasibility of the platform by carrying out the demonstration application on typical subject.

From 2020 to 2030: We should finish the construction of the large-scale professional database in whole country, realize the agriculture information service network around China; make breakthroughs in small scale nutrient information acquisition technology, remote sensing of crop quality monitoring, the growth simulation of animals and plants, variable fertilize equipment; realize quantitative management of soil, water and weather resources; realize dynamic monitoring of farmland water, pest and weed, realize intelligent precise management of farmland water, pesticide, cultivation and breeding; establish the network platform for agricultural virtual research to provide high-power service and cooperation environment for evolving a series of cross-regions, cross-organizations and cross-subjects agricultural scientific research activities in many fields, such as animal and plant breeding, high-yield cultivation, forecast and monitoring of pathogen and pest, the monitoring and assess of soil quality and weather forecast.

From 2030 to 2050: We should finish the construction of varies scale information collection system and realize the agricultural resources digital management basically; complete the development of a series of intelligent equipment and the precise management to animals and plants will be conducted during the production process; realize comprehensively agricultural informationization and precise management; realize the wide application of the network platform for agricultural virtual research, system optimization, and their systemic evaluation.

7.3 Technological Development Roadmap

7.3.1 Scientific Mission

The scientific mission of agricultural information and precise agriculture in future include: The key theoretical technology and system integration in agricultural information service network; The key theoretical technology and system integration in digital management of agricultural resource; The key theoretical technology and system integration in precise management during agricultural production process; The key theoretical technology and relative products of intelligent agricultural equipment; The construction of the network platform for agricultural virtual research and the conduction of relative key research projects.

7.3.2 Design Ideas for Roadmap

The agriculture information service network mainly depends on the network communication system, which application effect is remarkable and should give priority to make progress before 2020. The basic design idea is that "road-car-goods". Taking the national internet, the wireless communications network and the cable television network as the information superhighways (road), constructing information standardization and resource sharing agricultural information collective services platform (car), establishing information system, network information search system and analysis forecast system and so on (goods).

Digital management of agricultural resource, precise management of agricultural production process, intelligent agricultural equipment depend on many high-tech development, which are still at the research period of exploitation, need to plan with long-term development. The basic design idea is breaking through the key technologies difficulty first, solving the bottleneck problem, then studying and developing the key component, carrying on the important system's integration finally.

The construction of the network platform for agricultural virtual research is a huge systemic project, which needs to appraise the existing network information service project and the agricultural scientific research informationization resources first, then track and integrate next generation internet technology, and design the system structure of the platform, finally obtain the platform system prototype. On this basis, through establishing certain typical special the virtual research network platform to carry on the application and demonstration, explorating agricultural scientific research informationization environmental construction experience, we can construct the network platform for agricultural virtual research in China finally.

7.3.3 Roadmap

1. The key technologies and system integration of agricultural information service network

Such technology is relatively mature, we can basically achieve the goal by 2020. Its technical roadmap is shown as follows: using the national information highway network, based on national, provincial, municipal and county, four levels of agricultural information network center construction, upgrade, replacement, the common software support platform for agriculture information and national agricultural information resource base will be established, and the modern rural information service system with power function should be also developed. We will research and develop of professional systems, especially the technical model of agricultural strategy, production process and product flow based on Web GIS, agricultural recourses and macro strategy system, production process management system, agricultural products market analysis and distribution management system, develop the agriculture-specific search tools and navigation tools that improve the relevance and efficiency of agricultural information service. We will also establish large professional database, especially the recourse archive database, given the basis of resource optimization and precise management.

We should concentrate the power to overcome some key technologies, such as important mechanistic models for animal or plant growth, farmland ecosystem information and energy flow models, major agricultural pests and livestock disease monitoring and early warning control system, and Ipv6 technology and grid technology based agricultural applications.

2. Key theoretical technology and system integration of digital agricultural resource management

Real-time information collection, which is difficult to study, is the heart of agricultural information and precise agriculture. No real-time data updates, no precise management of resources. Currently, crop growth, yield, drought monitoring remote sensing and soil moisture monitoring technology are relatively mature. While resource management data yet will be realized after the break of nutrients, quality and other monitoring technologies. Therefore, three stages should be followed, the technological roadmap (Fig. 7.1) will be: ① From 2008 to 2020: we should improve yield estimation, crop growing, drought, and (weather, pests) disaster monitoring system, integrating soil moisture monitoring system, Overcome the key technologies of nutrients, quality monitoring based on optical, acoustic, magnetic and sensors technology. ② From 2020 to 2030: we should establish and improve the crop quality monitoring system and mobile field information collection equipment development and system construction based on remote sensing technology, complete stationary field information collection equipment development and system construction based on sensor networks, try to realize digital agricultural resources and digital growth process. ③ From 2030 to 2050: we should complete the digital national agricultural resources, access to the information service network, and provide digital services.

3. Key theoretical technology and system integration of precise management for agricultural production process

It is the mainly reflection of the production process precise management, which is necessary to be integration based on digital research and agriculture intelligent equipment research. The technology roadmap is as follows: ① From 2008 to 2020: digital model of plants or animals growth and software platforms should be created, to accomplish digital simulation and digital design; major crops digital production technology system should be constructed based on 3S technology; livestock, aquaculture digital farming technology platform and animal RFID tracking system should be established, to realized digital management of individual and groups. Computer-aided design system for plant production factory will be studied and developed. ② From 2020 to 2030: some quick access technologies and equipment will be investigated and exploited, which could be used to obtain the information of soil water, nutrient and crop growing status, yield, quality, plant diseases, insect pests, and weeds; the key technologies and equipment of digital production for main crop will be investigated, some machine will be manufactured and used in on-line measurement of crop yield, precise fertilization, precise seeding and employing of pesticide; developing digital sensor and intelligent control systemic information of biology and environment, establishing digital agricultural production factory, and realizing digital control and management during the process from seeding to harvest. ③ From 2030 to 2050: production process will be managed, practiced and spread precisely, and then high yield, resource efficiency and high economy benefit will be also realized.

4. The key theory, technology and relative production for agricultural machines and intelligent equipment

Some key developable equipments should be emphasized, which are the hardware support of the precise agriculture, including integrated techniques realized mechanization during the all produce processes of main crop, integrated equipments used for breed aquatics, technical equipment for facility agriculture, and new efficient intelligent technical equipment. We should centralize the power to create the production technology and materials for the key parts of the complete sets in agricultural machines, to plant precisely, to produce key technology and equipment for 3S position, to manufacture the production technology and materials for precise fertilize and pesticide, to make the technology and key equipment for environmental monitoring in facility agriculture, to produce the key equipment and materials for efficient saving on water irrigation, to create key technology for relative intelligent equipment and agricultural robot. Entire mechanization and intelligent during the agricultural production will be realized through the system integration. The main technical routes are presented as follows: from 2008 to 2020, a series of equipments should be manufactured, including automatic navigation control equipment in tractor, variable machine in chemical fertilizer, intelligent efficiency machine in pesticide spraying, efficient planter, seeding germination equipment, automatic equipment for plant grafting, transplanting and planting, intelligent precise technology and productions to monitor and control the environment during the agricultural production. In addition, these equipments should be entered the domestic and foreign markets. From 2020 to 2030, key parts of agricultural robot should be developed, and series of whole products will be created, and then demonstration and stereotypes of these products will be realized. From 2030 to 2050, the main products of the technical level and manufacturing level will close to the level of internationally renowned companies, some of the equipment will reach the international advanced level, and some international brands will be cultivate. A series of agricultural robots will be manufactured and will be applied large-scale in practice.

5. The construction of network platform for virtual agriculture research

In future, it is the basis of the large-scale agricultural research and innovation, which including: system structural design and model construction in network platform of virtual agriculture, management and sharing mechanisms of agricultural data resources and research equipment, key technologies of environmental service for agricultural information sciences, typical special of agricultural research and demonstration project activities, such as construction of virtual platform for plant breeding and animal epidemic prevention and control. The main technical roadmap is presented as follows: from 2008 to 2020, the network platform system construction will be analyzed and platform service system will be evaluated, the prototype of the network platform will be established. At the same time, the key technologies in agricultural Information service will be studied, which include integrated application of data collection equipment and technical criterions in data transfer; the sharing technical research on the resources of agricultural data, calculation and basic facilities should be finished; the key technical of data analysis and support environment in research activities of information agriculture will be completed, the various functions of the network platform for virtual research in agriculture will be realized initially, on this basis, two or three typical special of agricultural demonstration project activities will be carried out. From 2020 to 2030, the network platform for virtual research in agriculture will be applied in large-scale and further optimized in system, and the actual effect of the platform will be evaluated systemically. From 2030 to 2050, innovation agricultural research will be constructed through the network platform of virtual agriculture.

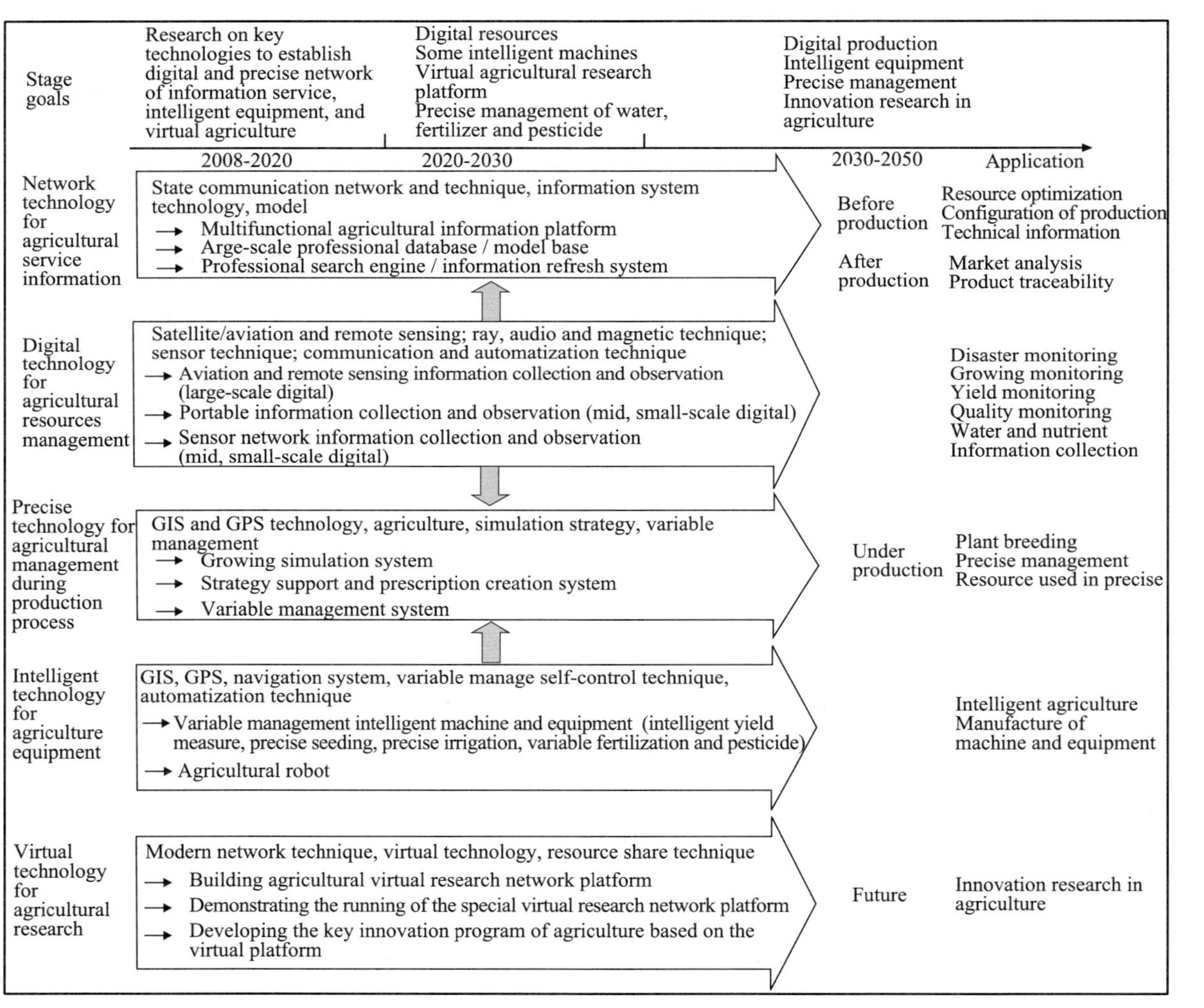

Fig. 7.1 The general roadmap of agricultural modernization and intelligentization science and technology development

8 The Institutions and Policy Support for Agricultural Science and Technology Development in the Future

It is urgent for China to introduce a series of institutional and policy supports to realize the goals of the roadmap for agricultural science and technology by 2050. To enable the agricultural science and technology in China to develop smoothly in line with the proposed roadmap and achieve the expected results in the five major areas of science and technology: plant germplasm resources and modern breeding, animal germplasm and modern breeding, saving resource uses, agriculture production and food safety, and modernization and intelligentization. To realize the overall goals of the roadmap for agricultural science and technology by 2050, it is urgent for China to establish a new agricultural science and technology innovation system, increase investment in agricultural science and technology, improve the investment priority in agricultural science and technology, and cultivate a large number of talented and innovative teams to enhance the autonomy of agricultural science and technology innovation. Specifically, the existing institutional and policy system of agricultural science and technology development needs reform, adjustment and improvement in the following areas.

8.1 Deepen the Reform of Agricultural Science and Technology, Establishing a National Innovation System for Agricultural Science and Technology

First of all, improving and perfect the national agricultural science and technology innovation system, while strengthening intellectual property protection, define the main changes in the roadmap of agricultural science and technology innovation system in various stages.

A series of studies at home and abroad show that agricultural science and technology has the natures of both public and private goods. The majority of agricultural science and technology shares of public welfare. The protection of many intellectual property rights of agricultural science and technology is difficult [1,2], especially in the developing countries like China. Therefore, it is difficult for enterprises or private sector to get access to all areas of agricultural science and technology in large-scale in the short term (or before 2020), particularly for parts of technologies in the areas involved in this report, such as the conservation of germplasm resources in "plant germplasm resources and modern breeding science and technology" and "animal germplasm and modern breeding science and technology", all the science and technologies related to environment in "resource saving science and technology", the majority of the core science and technologies in "agriculture production and food safety science and technology", and the digital management of agricultural information and resources in "agriculture modernization and intelligentization". Because these sciences and technologies are closely related to the resource, environment, national security, and strong externalities, the investment from private sector in these areas is difficult to reach the optimal level [3, 4]. Meanwhile, the national conditions in China determine that the agricultural science and technology is more of public domain because China has not established a good system for intellectual property rights (IPR) protection and also lacks a powerful agricultural research enterprise [2].

Of course, agricultural science and technology is more of public welfare but that does not mean that the private sector cannot be one of the mainstays of agricultural science and technology innovation system. Since the 1980s, the investment from private sector in agricultural research has kept up an impressive speed of rapid growth. In developed countries, private investment in agricultural science and technology has accounted for about 50% [1,3-5]. Meanwhile, the development of the private agriculture research system and investment significantly affects the investment and development of the public sector. In the United States and other developed countries, the agricultural research and development activities and investment from private sector has exceeded the governmental public sector in many areas. But in China, by the late 20th century, the investment in agricultural research and development from private sector has accounted for less than 2% of the total investment[2]. However, the situation has improved recently. Since 2000, the investment in agricultural research and development from private sector in China has increased at an average annual rate of 27%. By 2006, total investment from private sector has reached 22% of the 14.7 billion public investments [6].

An inadequate IPR protection system is a major factor that limits the growth of private investment[2,4,5]. Laws and regulations for IPR protection have been established gradually since the late 1990s in China, and the Chinese government has also launched a new round of innovations in the agricultural research system, strengthening IPR protection. However, intellectual property rights for agricultural science and technology in many areas cannot be protected effectively in a long period of time, and it will be a fairly long process for China to establish a good IPR protection system of agricultural science and technology.

At the same time, many studies also show that private sector investment in agricultural research and development not only focuses on the technologies whose intellectual property rights is easy to be protected, but also with high selectivity. In general, agricultural research and development investment from private sector is mainly concentrated in the applied and basic-applied areas. Agricultural research and development investment from private sector is also impacted evidently by the investment of public sector. Governmental research and development investment in basic research played a positive role in promoting the research and development investment from private sector in the applied and basic-applied areas, while governmental research and development investment in the applied and basic-applied areas has an opposite effect on the research and development investment from private sector [2, 4]. The study on the research and development investment from private sector also shows that in early 21st century, the research and development investment from private sector mainly concentrated in the areas whose intellectual property is easy to be protected such as agricultural processing, feed industry, animal and plant breeding etc., and the research and development investment in agricultural processing and animal husbandry and veterinary medicine company alone accounted for 70% or more of the total research and development investment from private sector in China. These studies show that on the one hand, the government should strengthen IPR protection to fully mobilize the enthusiasm of private sector to participate in agricultural research and development. On the other hand, the government should also gradually shift its focus of agricultural research and development investment into the areas of public goods (such as basic and basic-applied research) and those agricultural science and technology areas whose IPRs are not easy to be protected to promote the establishment of national agricultural science and technology innovation system.

To this end, we propose the following roadmap for national agricultural innovation system: by 2020, China will continue to strengthen the public functions and the dominant position of public service for agricultural science and technology and continue to create the national agricultural science and technology innovation system whose main body is the public sector; meanwhile, China should strengthen intellectual property rights protection to create a favorable investment and market environment for private to increase their enthusiasm in agricultural research and development investment and gradually cultivate modern large-scale agricultural research and development enterprises. By 2030, China should focus on strengthening the public functions of agricultural public sector of science and technology and put the science and technology focus of public sector on agricultural science and technology and the related basic and basic-applied research fields. At the same time, the intellectual property rights of agricultural science and technology will get full protection and the modern agricultural research enterprises or corporations will become the important constituent part of national agricultural innovation system. From 2030 to 2050, with full protection of IPR, a national modern agricultural science and technology innovation system will be established, where modern agricultural research corporations will play the critical role, and public and private sectors will be complementary in agricultural research and development.

Second, speed up the introduction of supporting policy for agricultural science and technology system innovation, and establish an effective incentive system for scientists.

Practices and previous researches have shown that it is hard for the science and technology sector and the agricultural sector to solve by themselves in many of the issues and challenges in agricultural R&D system, incentive mechanism and the investment mechanism. Reform involves a series of policies, including the social security system, personnel management, financing, wage and other incentive, and so on. For example, the delay in social security system reform and the lack of adequate funding for reform, among many others, are the major causes for difficulty to reform China's current agricultural science and technology system [7]. One of the most important reasons for the difficulties in agricultural research reform is the lack of adequate investment in reforms that causes the difficulties in personnel separation, formation of excellent research groups, establishment of an effective incentive mechanism in scientific research institutions and deepening reform of other issues in agricultural research.

The current agricultural science and technology innovation system lacks an effective incentive mechanism. In the agricultural research sector, a series of problems remain to be further studied and resolved in the performance evaluation criteria, research staff recruitment and dismissal, the relationship between the general treatment of public interest research scientists and income of other competitive units. In agricultural technology extension, the incomes of extension staff seriously decouple from the their performance, which has discouraged their engagements in technology extension.

It is recommended that a reform leading group that consist of the leaders from major ministries should be established. This will ensure the smooth progress and successful implementation of agricultural research and extension reform. The reform should also try to improve incentive of scientists and extension staff with better performance evaluation system.

To smoothly achieve the overall goal of the roadmap for agricultural science and technology development to 2050, it is necessary for China to make major adjustments on the future agricultural development planning. In recent years, China has announced a series of long-term strategies and planning for agricultural science and technology development, making a full deployment for the agricultural science and technology development after 2015 or even 2020. However, considering the rapid development of global technology and the changing demand for agricultural science and technology, the future agricultural science and technology development planning will have to adapt to these changes in time and make the appropriate adjustments. To this end, it is proposed to adjust the future agricultural science and technology development planning in accordance with the general goals, stage goals and the corresponding key technologies referred in "Strategic Study of the roadmap for agricultural science and technology development to 2050."

In order to achieve the overall goal of the roadmap for agricultural science and technology development to 2050 with high efficiency, it is recommended to establish a new National Agricultural Science and Technology Committee, mainly responsible for the determination and timely adjustment of the roadmap for agricultural science and technology development to guide and supervise the achievement of general goals and stage goals in according to the global technology trends and national demands. Because there has not been a coordinating body in China so far to coordinate the national major programs and provincial projects, and the projects between provinces, the establishment of the National Agricultural Science and Technology Committee can also avoid duplication in project settings to a certain extent. In this case, the phenomenon of over decentralizing of research institutions and great duplication of research projects in agricultural research system are still widespread. The survey of researchers engaged in the research of main crops shows that, in recent years, with the continual increase of central and provincial investment in agricultural research, the phenomenon of applying the same research project from different departments and conducting similar research in different provinces are still quite wide spread, and the central and local, provincial and provincial overlapping projects problems have not been effectively addressed.

To ensure the realization of the roadmap for five major areas of agricultural science and technology to 2050 and the overall roadmap for agricultural science and technology development to 2050, it is proposed that the National Agricultural Science and Technology Committee should be made up of the main ministries, Chinese Academy of Science and the provincial agricultural departments.

8.2 Establish National Agricultural Long-term Technology Investment Mechanism and Improve the Investment Structure of Agricultural Science and Technology

First, establish a national agricultural long-term technology investment mechanism to increase investment in agricultural science and technology.

Establish a long-term increase mechanism for agricultural scientific research investment as soon as possible to ensure the stable increase of the national agricultural research investment, especially supporting the long-term and basic researches. It is recommended that the public agricultural research investment intensity (ratio of research investment and agricultural GDP) should be increased from the 0.47% in 2005 to 1% in 2015, 1.5% in 2020, and maintain at about 2% in 2030-2050. Meanwhile, strengthen the implementation of IPR protection, improve patent protection system, provide preferential policy, provide incentive mechanism for private sector to engage in agricultural research.

Second, improve agricultural science and technology investment structure by raising the share of core funding in total investment.

Adjust the government's investment channels to raise the proportion of core funding (compared to competitive project funding) in total investment in research and provide a good research environment for breakthroughs of major sciences and technologies. Strengthen the science and technology, infrastructure and capacity building of the five major areas in the roadmap for agricultural science and technology to 2050, improve the conditions of the relevant scientific research, and increase China's capacity for independent innovation in agricultural science research to ensure the smooth implementation of the roadmap.

8.3 Train a Large Number of Talents and Innovative Teams, and Enhance the Ability of Independent Innovation of Agricultural Science and Technology

Establish National Agricultural Science and Technology Talent Foundation and focus on the attracting and training of outstanding leading scientists to engage in agricultural research. To ensure the smooth implementation of the roadmap, it is recommended that the national policies should strongly support a number of world-class scientific research institutions and universities, strengthen their research capacity building and improve capacity of independent innovation in the five major areas in the roadmap for agricultural science and technology development. It is also recommended that China needs to form a large number of stable, efficient, and innovated agricultural research teams.

Main References

[1] Alston JM., Craig BJ, Roseboom J. "Financing Agricultural Research: International Investment Patterns and Policy Perspective" . World Development, 1998, 26(6):1057-1072.

[2] Huang JK, Hu RF, Rozelle S. Agricultural Research Investment in China: Chanllenges and Outlook. Beijing: China Financial & Economic Press. 2003.

[3] Keith F. Trends in Agricultural Research Expenditures in the United States, Public-Private Collaboration in Agricultural Research. Ames: Iowa State University Press, 2000

[4] Carl PE. 'The Growing Role of the Private Sector in Agricultural Research' , Agricultural Research Policy in an Era of Privatization. Walling ford: CABI Publishing. 2002: 35-50.

[5] Hu RF, Liang Q, Huang JK, et al. Priveate Agricultural Research Investments in China: Current Situation and Past Trend. The research report of CCAP, Chinese Academy of Sciences, 2009.

[6] Huang JK, Hu RF. To Perfect the Agricultural Research Revolution and Promote the Agricultural Research Innovation. The briefing of CCAP, Chinese Academy of Sciences, 2007(2).

Epilogue

Seeing the daunting subject "Roadmap for Agricultural Science and Technology Development in China to 2050" of this report, it can be difficult to put pen to paper. After all, this report is about circumstances in forty or fifty years. It is no exaggeration to say that taking over the research work means to treat it with awe just like you are walking on thin ice. Nevertheless, this report is a very serious study of the roadmap for agricultural science and technology development. Such feelings of fear and trepidation may not cease in the future because the future means that the report will be constantly faced with practice tests and criticism.

Needless to say, it is of important research value for the Chinese Academy of Sciences to study and formulate this report, "Roadmap for Agricultural Science and Technology Development in China to 2050," which has important strategic significance to support Chinese agriculture in a new era of global agricultural development. This report studies the technological development roadmap of plant germplasm resources and modern breeding technology, animal germplasm and modern breeding, agricultural science and technology based on resource frugality, science and technology of agriculture production and food safety, and science and technology of agriculture with modernization and intelligentization. The report seeks to give technical support for the sustainable development of science and technology in China to 2050 under the pressures and challenges from resources and environment. Specific areas include increasing the production potential of plants and animals, resource management, utilization and security, agricultural production and food security, and ecological and environmental protection.

The report "Roadmap for Agricultural Science and Technology Development in China to 2050" is comprehensive, complex, and a very rigorous study of agricultural science and technology development strategy. Because it studies plants, animals, resources, security, modern agricultural, institutional policy and other areas, it is clear that neither the person in charge nor a small number of experts could complete the report. The report is our collective results accomplished jointly by study group members and relevant experts bringing about collective wisdom. To complete the research, the members of the study group should not only have long-term accumulation of related academic knowledge and wise consideration on the cutting-edge issues and major trends, but they should also have the courage and boldness to accept criticism. With the concerted efforts and co-operation of the study group members and relevant experts, "Roadmap for Agricultural Science and Technology Development in China to 2050" was finally finalized and put into publication in one and a half years. All members of our group are greatly delighted and feel the enormous pressure at the same time. This is because the Roadmap for Agricultural Science and Technology Development depicts the blueprint for the development of agricultural science and technology in forty or fifty years. Although the group members have all immersed themselves in their respective areas for a long time with a modicum of success and have made painstaking efforts in this report, in face of the unpredictable future, we are acutely aware that human cognition is limited after all. The study has been modified again and again, demonstrated repeatedly, but even so, omissions and mistakes may still occur; therefore, we believe that in the future, as agricultural science and technology develop, it is necessary to publish a "Roadmap for Agricultural Science and Technology Development" every 5–10 years, which would play an important role in promoting socio-economy and agricultural development in China.

Here, we first want to thank the experts and scholars participating in this report's discussions, brainstorming, writing, and revision. They are: Jun Xia, Yongbiao Xue, Fangzhen Sun, Kang Zhong, Ming Dong, Shiping Tian, Zhijie Wu, Lisong Dong, Ping Xie, Guofan Zhang, Qi Zhou, Feng Ge, Zhiliang Tan, Yanhong Wang, DongMei Zhou, Dr. Yiping Zhu, and Dr. Yuan Yao. We will specially give our thanks to Zhensheng Li, Zhizhen Wang, Shengli Yang, Xiaosheng Chen, Jiayang Li and Director General Zhibin Zhang, who participated in the guidance and assessment of the report, and Honglie Sun, Zhihong Xu, Zuoyan Zhu, Yiyu Chen and Vice Minister Taolin Zhang for peer review of the final draft of the report and the valuable advice. Our report would not have been perfected without their guidance and help. Finally, we want to give thanks to the 20 experts in the agricultural strategy research group for remaining engaged in the discussion and preparation of the report. Thank you all for collaboration! We are lucky at least to have the opportunity to participate in such a collective research work, to have this opportunity to share with friends in different areas, and to discuss or even argue, whatever the future will outcome!

Finally, we would like to say that scientists are not prophetic, "Roadmap for Agricultural Science and Technology Development in China to 2050" is a subjective judgment and describes the blueprint for agricultural science and technology development in forty or fifty years, formulated by the Research Group for Agricultural Strategy in Chinese Academy of Sciences according to the domestic and overseas demand for agricultural development and the development trend of technology. Therefore, whether readers regard the blueprint as a bold guess of scientists or as our common good wishes, if this book has brought about some enlightenments for you, all the efforts of the members of the task group will be considered worth while!

Research Group on Agriculture of the
Chinese Academy of Sciences

January, 2011